Microcomputer-Based Input-Output Modeling

Microcomputer-Based Input-Output Modeling

Applications to Economic Development

EDITED BY

Daniel M. Otto
and Thomas G. Johnson

LONDON AND NEW YORK

First published 1993 by Westview Press

Published 2018 by Routledge
52 Vanderbilt Avenue, New York, NY 10017
2 Park Square, Milton Park, Abingdon, Oxon OX14 4RN

Routledge is an imprint of the Taylor & Francis Group, an informa business

Library of Congress Cataloging-in-Publication Data
Microcomputer-based input-output modeling: applications to
economic development / edited by Daniel M. Otto and Thomas G. Johnson.
p. cm.
ISBN 0-8133-1046-6
1. Input-output analysis—Data processing. 2. Economic
development—Mathematical models—Data processing.
3. Microcomputers. I. Otto, Daniel. II. Johnson, Thomas G.
HB 142.M535 1993
339 .2'3'0285416—dc20 89-78298
CIP

ISBN 13: 978-0-367-00393-7 (hbk)

ISBN 13: 978-0-367-15380-9 (pbk)

Contents

Preface

This book brings together papers by researchers and analysts from academic, governmental, and other agencies that discuss recent developments in I/O models for microcomputers, applications of I/O models in regional studies, and to explore future directions for the methodology. Interest in these topics has grown in recent years as the availability of expanded PC capability and the adaptation of I/O software and data by academic and for-profit organizations has increased the set of modeling options for individuals and organizations interested in using I/O analysis for applied research and policy analysis. As enhanced PC capacities and software developments expand the pool of potential I/O users, the possibilities for misapplication also increase. The proceedings from this conference are intended to provide reference materials on I/O applications for this evolving audience of new I/O users.

The objective of illustrating different I/O applications ranges from the traditional to more sophisticated extensions of the basic I/O framework. These applications include standard impact analysis and structural analysis of a regional economy, as well as extending the I/O framework to dynamic and simulation applications and development of Social Accounting Matrices (SAMs) and Computable General Equilibrium (CGE) models. In addition to these applications, the book provides a review of I/O methodology and different methods of constructing I/O models, a discussion of appropriate and inappropriate uses of I/O methodology, and a discussion of alternative sources for developing or acquiring an I/O model.

This range of topics is intended to appeal to the newly initiated as well as the more experienced I/O users interested in more advanced applications of the I/O framework. This grouping of topics also becomes the organizational scheme for this book.

The opening section of the book provides background information on traditional I/O models, including chapters on I/O theory and methodology, chapters on survey approach and secondary data approaches to developing I/O models, cautions on the use of I/O models in applied work, and considerations in choosing an I/O based modeling system. The material in this section is intended to provide relatively inexperienced users with sufficient background or reference material to begin conducting applied I/O work on a number of topics.

The second section of the book focuses on presenting a set of I/O applications illustrating a range of potential uses for an I/O modeling system. The set of applications illustrated in this section includes a structural evaluation of the farm sector in the economy, the impact of the Conservation Reserve Program, the distributional impacts of social security programs, and the use of an I/O model to help regions target industrial recruitment candidates. This section is intended to provide a set of working examples of I/O applications for users to draw upon for reference in analyzing similar issues.

The third and final section of the book continues the emphasis on applications as the discussion shifts to extending the original I/O model to several more innovative and sophisticated techniques. The first two chapters in this section discuss application of I/O techniques as a core component in simulation modeling systems for public service planning activities and policy analysis. The use of I/O components to develop and apply more sophisticated analytical techniques of dynamic I/O modeling, SAMs, and CGE models are discussed in three other chapters in this third section. The topics of SAMs and CGE modeling have a large and evolving literature on their development and applications. The discussion of SAMs and CGE models are intended to describe the basic components of these modeling systems and their relationship to the basic I/O model with a discussion of potential applications of these techniques.

Finally, thanks are in order to the North Central Regional Center for Rural Development for its generous support in the planning and implementation stages.

Daniel M. Otto

About the Contributors

Michael W. Babcock, Professor, Department of Economics, Kansas State University, Manhattan, Kansas.

G. Andrew Bernat, Jr., Economist, United States Department of Agriculture, Economic Research Service, Washington, DC.

Curtis H. Braschler, Professor, Department of Agriculture Economics, University of Missouri-Columbia, Columbia, Missouri.

Sharon M. Brucker, Research Associate, Department of Food and Resource Economics, University of Delaware, Newark, Delaware.

Stephen C. Cooke, Assistant Professor, Department of Agriculture Economics, University of Idaho, Moscow, Idaho.

Randal C. Coon, Research Specialist, Department of Agriculture Economics, North Dakota State University, Fargo, North Dakota.

Gary T. Devino, Professor, Department of Agriculture Economics, University of Missouri-Columbia, Columbia, Missouri.

Gerald A. Doeksen, Professor, Department of Agriculture Economics, Oklahoma State University, Stillwater, Oklahoma.

Brenda L. Ekstrom, Research Associate, Department of Agriculture Economics, North Dakota State University, Fargo, North Dakota.

Frank M. Goode, Associate Professor, Department of Agriculture Economics, Penn State University, University Park, Pennsylvania.

Thomas R. Harris, Professor, Department of Agriculture Economics, University of Nevada, Reno, Nevada.

Steven E. Hastings, Professor, Department of Food and Resource Economics, University of Delaware, Newark, Delaware.

Mark S. Henry, Professor, Department of Agriculture Economics, Clemson University, Clemson, South Carolina.

David Holland, Professor, Department of Agriculture Economics, Washington State University, Pullman, Washington.

Thomas G. Johnson, Professor, Department of Agriculture Economics, VPI & SU, Blacksburg, Virginia.

David S. Kraybill, Assistant Professor, Department of Agriculture Economics, Ohio State University, Columbus, Ohio.

F. Larry Leistritz, Professor, Department of Agriculture Economics, North Dakota State University, Fargo, North Dakota.

Jay A. Leitch, Professor, Department of Agriculture Economics, North Dakota State University, Fargo, North Dakota.

Timothy L. Mortensen, Research Assistant, Department of Agriculture Economics, North Dakota State University, Fargo, North Dakota.

Gerald Schluter, Agricultural Economist, Agriculture and Rural Economy Division, Economic Research Service, Washington, DC.

George I. Treyz, Professor, Economics Department, University of Massachusetts, Amherst, Massachusetts, and President of Regional Economic Models, Inc. - (REMI).

Mike D. Woods, Professor, Department of Agriculture Economics, Oklahoma State University, Stillwater, Oklahoma.

Peter Wyeth, Associate Professor, Agriculture Economics and Marketing, Washington State University, Pullman, Washington.

1

An Introduction to Regional Input-Output Analysis

Steven E. Hastings and Sharon M. Brucker

Introduction

Sound economic development decisions require information about the impacts of economic growth and/or decline and the relative benefits and costs of alternative development strategies. Increasingly, university researchers, Cooperative Extension personnel and state and local development officials are being called upon to provide this information. This is as true in the many communities experiencing economic growth as in communities facing stagnation or economic decline. In the former group, the problem is one of managing the growth, e.g., what impact would growth in the banking industry have on the housing industry or transportation system, or what impact would growth in the summer home market have on the agricultural sector? In the latter group, the question is how to stimulate economic growth. Typical issues would be: what will be the impact of a manufacturing plant closure or what resources does the community have to offer to potential industries seeking a plant location?

Attempts to answer these and many similar questions have led to the continued development and increased use of complex analytic models to assess economic impacts; computerized input-output, industrial location, demographic and fiscal impact models are readily available to the economic analyst. However, the use of these models must be accompanied by a concomitant familiarization with the potential uses, underlying assumptions, data requirements, weaknesses and strengths of the models. Without this familiarity many analytic models and the information they provide will be unused, underused, or misunderstood.

While there are many methods of regional analysis (Isard, 1960) (Richardson, 1972), there has been renewed interest in and use of regional input-output in recent years.[1]

The purpose of this chapter is to familiarize the reader with the theoretical framework, construction and use of regional input-output models. The description of the analytical framework of an input-output model includes a discussion of the components of the model, a section on analytic measures derived from the model, and a section on the assumptions of the model. A final section presents the phases of model planning, construction and use, including some of the inherent limitations and problems. Finally, some suggestions for effective use of the model will be provided.

Input-output analysis is frequently chosen for regional analysis because it provides several types of information. It is an excellent descriptive tool, showing in detail the structure of an existing regional economy. It provides important information on individual industrial sector size, behavior and interaction with the rest of the economy. It shows the relative importance of sectors in terms of their sales, wages, and employment. It also provides a way to predict how the economy will respond to exogenous changes or changes that are planned. Therefore, it is useful in prescriptive exercises where various actions are being considered and the relative merits are to be determined based on alternative outcomes.

The Basics of Input-Output Analysis

This section will present in nontechnical terms the basic formulation of an input-output model. Similar treatments can be found in Trench and Frick (1982), Bills and Barr (1968), and Brucker and Hastings (1984). Readers interested in a more technical discussion should see Miernyk (1965) or Miller and Blair (1985).

Input-output analysis attempts to quantify, at a point in time, the economic interdependencies in an economy, such as a nation or a state. In this analysis, all economic activity is assigned to one of two types of sectors: production or final demand. Production sectors (e.g., agriculture, manufacturing, trade) represent all establishments in the region producing a specific product or service. The output levels of production sectors are determined within the model and are therefore termed endogenous to the model. Sectors representing final demand may include households, government, foreign trade. The levels of activity in these sectors are assumed to be determined by forces outside the model, e.g., government policy, and are therefore termed exogenous. All changes in the endogenous sectors of a input-output model are results of changes in the exogenous sectors.

A fundamental underlying relationship of input-output analysis is that the amount of a product (good or service) produced by a given sector in the economy is determined by the amount of that product that is purchased by all the users of the product. The users include other industrial sectors that use the product as inputs in the production of their own products (collectively referred to as intermediate demand) as well as sectors that use the product in its final form (collectively referred to as final demand). As an example, the amount of refined petroleum produced is determined by the intermediate demand (oil for plastic products, fuel used by farm tractors) and the final demand (heating fuel for consumers, gasoline for consumers' cars).

The flow of products between sectors is measured in dollars and referred to as transactions between the various sectors. An important assumption of input-output analysis is that transactions (flow of products) between sectors is a fixed and constant proportion of the amount of product being produced. For example, the dollar amount of petroleum that the agricultural sector buys is a fixed proportion of the dollar amount of agricultural products produced.

Components of a Regional Input-Output Model

To use the input-output framework for regional analysis, it is necessary to establish an input-output model specific to the region. Three prescribed tables (or matrices) make up an input-output model: the transactions table, the direct requirements table and the total requirements table. Generally, the transactions table shows all the transactions between the various sectors in an economy. This provides a snapshot at a point in time of all the economic activity in the economy. These data can be used to produce a table of direct requirements which show how much of each input is required to produce one dollar of output. Using the direct requirements, a table of total (direct and indirect) requirements can be determined. These can be used to determine the impact on the entire economy of a change in any one sector or combination of sectors. The derivation and interpretation of these three tables are discussed below.

The Transaction Table

The transaction table is used to organize all the information about the regional economy; the other tables are derived from the transactions table. The transactions table shows the flow of all goods and services produced (or purchased) by sectors in the region (Table 1.1). The key to understanding this table is realizing that one firm's purchases are another firm's sales and that to produce more of one product requires the production of more of the inputs for that product. The transaction table can be created or read from two perspectives. Reading down a column gives the purchases by the given sector named at the top from each of the other sectors named at the left. Reading across a row gives the sales of the sector named at the left to each of the sectors named at the top.

In the hypothetical Region X depicted in Table 1.1, reading down the first column shows that agricultural establishments buy $202 worth of their inputs from other agricultural establishments. The sector also buys $34 worth from manufacturing firms, $47 from trade establishments, and $86 of services. Amounts are also given for a sector's expenditures on payments to non-processing sectors such as the government and household sectors which receive "payments" as taxes and wages, respectively. In Region X, agriculture purchases $200 of labor inputs and capital services from the household sector.[2] It is also important in a regional economy to keep track of how many of the inputs are purchased outside of the region. In this example, agriculture purchases $172 of its total inputs from outside the region; these are listed as imports.

The total expenditures made by the agriculture sector are $741. In an input-output model, total sales (output) and the total purchases (expenditures or inputs) are always equal for a given sector.[3] Reading across the first row shows that agriculture's sales are distributed as follows: $202 worth sold to agriculture, $182 worth to manufacturing, $10 sold to trade, and $47 worth to the service sector. The other $300 worth of agricultural products are sold to consumers or government establishments or exported out of the region. Sales to final demand sectors are shown as $100 to the households column (consumers purchases) and $200 in the "other" final demand (in some models this is broken down into further government sectors, inventories and exports).

The transactions table is important because it provides so much useful information about the economy of the region. Not only does it show the total output of each sector but also sectors upon which each sector is dependent. It also indicates the sectors from which the residents earn income.

To correctly understand a transactions table, it is important to be familiar with three conventions almost universally followed in constructing it. The transactions are denominated in producer prices; the price to the final users (collectively, the value of final sales) is accounted for only after the expenditures for transportation and trade sectors are included. The output of these two sectors, transportation and trade, is usually denominated in value added (i.e. sales net of the value of the merchandise purchased for resale, or for transport). For example, if an individual purchases a car at a retail dealership for $10,000, this is not treated as a $10,000 transaction with the retail sector. Instead, this flow is divided into a flow to the retail sector ($2,000), a flow to the services (transportation) sector ($1,000), and a flow to either manufacturing or imports ($7,000) depending on whether the car is produced in the region or not. This often necessitates adjustments of trade sales figures gathered from other data sources. It is also important to note how secondary products are handled, usually their value is included with the sector which produces them as a primary product.

Finally input-output transactions represent only the portion of the economy which is on "current account." Hence no investment activity, or capital account activities, are included in the analysis. This practice simplifies modeling by assuming that supply is infinitely elastic. This is a static equilibrium model and therefore capacity is assumed to be appropriate.

The Direct Requirements Table

Because the transaction table is in absolute dollar figures it is not easily generalized. However, important production relationships can be identified if the patterns of expenditure are stated in terms of the portion of total expenditures made by a sector for various inputs. This is done by dividing the dollar value of inputs purchased from other sectors by the total output (or expenditures). Each transaction in a column is divided by the column sum. The resulting table is called a direct requirements table (Table 1.2).[4]

The direct requirements table is similar to the transactions table in its format. Rather than showing the actual dollar transactions, the direct requirements table shows, for the sector named at the top, what fraction of total expenditures was made to purchase inputs from the sector named at the left. In Region X, agriculture purchases 27 percent (202/741*100) of its inputs from itself, five percent (34/741*100) from manufacturing, six percent (47/741*100) from trade, etc.

In the direct requirements table, each cell represents the dollar amount of that particular good or service *required from* the industry named at the left to produce one dollar's worth of the good named at the top of the column. Each column is essentially a "production recipe" for that industry's output produced for final demand. In Region X, for every new dollar's worth of agricultural output for final demand, agriculture will buy 27 cents worth of manufactured goods. If the final demand for agricultural products increased by $100,000 (for example, exports increased) then agriculture would increase purchases from manufacturing by $27,000. The same interpretation is given to other cells. It is important to note that in input-output analysis, this production formula, or technically, the column of direct requirement coefficients, is assumed to be constant and the average of all establishments within a sector regardless of input prices or how much is being produced. Furthermore, if the sector is aggregated to include production of many different products, the input needs for each product are assumed to be represented by the aggregated "average" direct requirements column.

It is critical to note that these are products and services required from establishments *in the region*. Thus, if the product or service needed is not available in the region, and must be imported, the direct requirements are not identical with a production function. Therefore, it might be clearer to refer to the table as a local requirements table, but the input-output literature has consistently used the term, direct requirements table.[5]

Knowing the production recipe for local products provides detailed information on how the local economy works and how the industries in it interact. Not only does it indicate the use of individual inputs, but also the total of inputs from other production and service sectors in the region. This total local purchases row is significant because it indicates the degree to which a sector will effect other industries in the region. Note that in Region X, manufacturing buys 73 percent of its inputs from other regional producers. This would be characteristic of a food processing industry in a farm state or any industry which relies on raw materials which are available in the region. The importance of this relationship with regard to regional impacts will be discussed later.

Assuming that the direct requirements will also represent spending patterns necessary for additional production activity, the effects on the regional economy of a change in final demand for the output of any one sector or group of sectors can be predicted. To illustrate, consider the effect on Region X of a $100,000 increase in export demand (final demand) for agricultural products. From Table 1.2, it can be seen that any new final demand for output from agriculture will require purchases from the other sectors in the amounts shown in the first column. These are known as the direct effects. However, in order for establishments in the manufacturing, trade and service sectors to supply inputs to the agricultural sector, they, in turn, will have to increase production. Each of these sectors will then require additional inputs as shown in each of their columns. These additional input requirements are known as indirect effects.

If agriculture increases exports by $100,000, it seems obvious that farmers will have to increase production by $100,000, but our production recipe shows that they will also have to supply $27,000 of farm products as inputs to the farmers producing the $100,000. This can be envisioned as grain to be used as feed for livestock. However, the impacts on local production do not stop here. Additional direct effects include $5,000 from manufacturing establishments (perhaps fertilizer or fuel for tractors), $6,000 from retail or wholesale outlets (probably representing seed, fertilizer and supply purchases), and $12,000 from the service sector (could be banking services, real estate or insurance expenses). Summing these purchases would give $150,000 of impact on the economy, the initial change ($100,000) plus the total direct effects ($50,000).

One of the strengths of an input-output model is that it does not stop at this point, but also measures additional or indirect effects of this increase in agricultural exports. It may be helpful to see how these indirect effects accumulate. Looking just at the indirect effects on the agricultural sector, in order for the farmers to produce the additional $27,000, they must have agricultural supplies of .27 for each dollar ($27,000 x .27 = $7,290).

Likewise, for the manufacturing firms to produce the fertilizer they will need some inputs from agriculture at a rate of .39 per dollar ($5,000 x.39 = $1,950). The direct requirement table shows that the trade establishments do not require inputs from agriculture, however, the service sector requires a .02 per dollar ($12,000 x .02 = $240). Summing these inputs required indirectly from agriculture gives $9,480 of additional production necessary ($7,290 + $1,950 + $240 = $9,480). In turn, these additions to production will require still more purchases from agriculture and from all of the sectors according to the recipe. Agriculture will need to supply (9,480 x .27) to itself and (2,220 x .39) to manufacturing and ($7,130 x.02) to services. This can be summed to be $3,569 ($2,560 + $866 + $143). Thus, an additional $3,569 of required inputs from agriculture alone will be added. If the initial change, the direct effects, and the first two "rounds" of indirect effects are summed, the required output equals $140,049. Thus, in Region X production required to meet the initial change in exports will be considerably greater than the $100,000 in agricultural products. Thus, from the direct requirements table and calculations of two rounds of effects, it can be seen that for every $1 of agricultural sales to exports (or other final demand), $1.40 dollars of agricultural production will be required.

The Total Requirements Table

Fortunately for analysts, the sum of these direct and indirect effects can be more efficiently calculated using matrix algebra. The methodology was developed by Wassily Leontief and is easily accomplished in computerized models. Typically, this result is presented as a total requirements table, also called the Leontief Inverse (Table 1.3).

Each cell in Table 3 indicates how much output from the sector named at the left will be required in total, as direct or indirect inputs, for a one dollar increase in final demand for the output from the sector named at the top of the column. For example, in the first row of the first column of Table 1.3, 1.42 (or $1.42) indicates the total output required by agriculture from itself in order to produce an additional dollars worth of agricultural output for final demand. The value is slightly larger than the 1.40 approached with the two rounds of calculations in the example above. If the total (direct and indirect) outputs required from each sector are added down a column, the total outputs required from the whole regional economy to produce an additional dollar of agriculture for final demand is 1.99. Thus, the interpretation of Table 1.3 is that the initial $100,000 increase in agricultural exports will generate production of $199,000 in the entire economy, $142,000 of which is in agricultural production, when all direct and indirect effects are considered.

Thus far, the increased output required has been governed by the production sectors in the model. However, additional production will also be required to meet purchases made by consumers who are spending the wages they were paid to produce the direct and indirect output discussed above. To estimate the magnitude of these "induced" effects, the household row and column must be included in the direct requirements table that is inverted. In essence, the household sector is no longer considered exogenous but is part of the endogenously determined portion of the model, i.e. the level of consumer demand is determined by the total amount of production in the regional economy and not by forces outside of the economy. When this is done, the model is a *closed model* and the total requirements table includes direct, indirect and induced effects (Table 1.4).

One assumption inherent in induced effects is that wage earners will spend their new income in the same pattern that they spent total earlier acquired income (i.e. follow the column of expenditures) and that all payments to households, such as dividends and profits, accrue to regional residents. Some models attempt to adjust for out-of-region household recipients. Table 1.4 indicates that when household expenditures and the additional production to meet these demands are included the total production required to meet the initial $100,000 export of agricultural products is $340,000. Most modelers warn that induced effects may be overstated because of the lack of validity of the assumptions above.

Multipliers -- Construction and Use

Of all the information input-output analysis can provide about a region, multipliers are the most commonly requested. Although multipliers have not yet been explicitly introduced, they appear above as the "total" requirements row in the total requirements table (Table 1.3 and 1.4). Each element in this row tells the analyst by what multiple each dollar of increased final demand will impact the overall output of the regional economy. Hence, it is referred to as an output multiplier. It measures how much total additional production of goods and services is required throughout the regional economy for every one dollar of additional final demand for the goods produced by the industry named at the top of the column. An output multiplier for an open model (direct and indirect effects) is a referred to as a Type I output multiplier and the multiplier from a closed model (direct plus indirect plus induced effects) is referred to as a Type II output multiplier.

In addition to output increases, users are frequently interested in knowing the impacts of changes in *final demand* on statewide income and employment. Because it is assumed that the ratio of income (or employment) to output given in the production recipe (i.e. household row of the direct requirements table) is fixed and will be the same for all changes in production, these ratios can be used with the total requirements table to determine the sectoral and total income and employment effects of any change in sales to final demand.[6] If a total output requirements coefficient (a cell in Tables 1.3 or 1.4) is multiplied by the income payment associated with each dollar of that output (household row, Table 1.2), the resultant coefficient will indicate how much income will be generated in the sector named at the left for a given dollar of output by the sector named at the top. Summing each column will provide the total direct and indirect (or direct, indirect and induced, if the closed model is used) income impacts.[7] These total income effects (for closed model, Table 1.4) are shown for Region X in Table 1.5. This table indicates that the initial 100,000 increase in final demand for agricultural products will lead to $40,200 (1.49 x .27) of increased incomes in agriculture, $900 (.11 x .08) in manufacturing incomes, $6,100 (.87 x .07) of income for trade, and 38,100 (.93 x .41) of income in the service sector. Therefore, the economy as a whole can expect $91,800 of increased income to be generated from the initial $100,000 of sales.

A similar procedure can be used to estimate region-wide employment effects. It is necessary to estimate a set of physical labor input requirements for all sectors by determining the ratio of employees in a sector to the total output of the sector. For example, if 221 people are employed in the agricultural sector, the physical labor requirement coefficient is .298 (221/741) per thousand dollars of output.[8] Multiplying each coefficient in the total requirements table by the sector's physical labor input requirement will provide a table of employment effects. Table 1.6 presents these employment effects (derived from Table 1.4).[9] Reading down the first column of Table 1.6 it can be seen that the initial $100,000 of sales to agricultural exports will generate a total of 123 jobs in the Region X's economy. The jobs will be distributed throughout the economy with 44 jobs (1.49 x .298) in agriculture, 2 (.11 x .21) in manufacturing, 32 (.87 x .36) in trade and 36 (.93 x .40) in service.

Frequently, analysts do not have data on initial final demand changes, but instead have data on changes in a sector's employment or income levels. With this information, alternative measures are needed to estimate impacts on the economy. These measures which relate sector changes in income or employment to total regional changes in income and employment are referred to as income and employment multipliers.

Income multipliers, can be estimated by calculating the ratio of the total income effect coefficient to the direct income effect. This multiplier could then be applied to expected new income to determine total income effects even if the expected change in final demand is not known. Such an income multiplier for an open model (direct and indirect impacts only) is called a Type I income multiplier and for the closed model (direct, indirect and induced effects) a Type II income multiplier. Likewise, Type I and II employment multipliers can be created using the direct physical input coefficient and the total employment effect coefficients.

It is generally acknowledged that although Type I multipliers understate the overall effects by ignoring wage-earner's increased spending, typically the Type

II multipliers overstate the impacts or if marginal expenditures are different than average expenditures.[10] Therefore, often the Type I and II output multipliers are used together to give a range of the expected impact. Appendix B presents a summary of the measures available from an input-output model for Region X and how they are derived.

Assumptions of Input-Output Analysis

When an input-output model and its associated multipliers is used for comparative static analyses, some rather rigid assumptions must be made about the nature of the production process. Several of these assumption have been alluded to previously; however, it may be useful to explicitly list the most significant of them (Bills and Barr, 1988; Miller and Blair, 1985):

(1) The output of each sector is produced with a unique set of inputs. (no substitution between inputs)

(2) The amount of input purchased by a sector is determined solely by its level of output (no price effects, changing technology, or economics of scale).

(3) No external economies of scale exist; the effect of additional types of production is additive only.

(4) The in-state and out-of-state distribution of purchases and sales is fixed.

(5) There are no constraints on resources (supply is infinite and perfectly elastic).

(6) Local resources are efficiently employed (no under employment of resources).

In short, this model assumes that market structure, state of technology, relative prices and geographic distribution of economic interaction are fixed and that the supply of inputs and demand for output are elastic.

Conducting Regional Input-Output Analysis

There are three major phases in conducting regional input-output analysis; planning, construction, and use. Even if a model is purchased from any of several on the "ready-made" market, the other two phases (planning and use) are still critically important for accurate analysis.

Planning

Several decisions will be made even before a model is purchased or constructed. It is preferable that they be made explicitly and simultaneously. They include; defining the region, identifying objectives of the study (includes identifying both impacts and sectors to be studied), and assessing the availability of the resource (human, data, computer hardware, and financial). A regional definition may need to be adjusted based on objectives and/or data availability. For example, a local community with a goal of a 20 percent increase in employment might be wise to define the region of study as something beyond existing city limits. Furthermore, if the objectives include impacts on economic variables or on small disaggregated sectors, the data may dictate a county or SMSA level definition.

Construction (Acquisition)

The region, the objectives, and the available resources will effect the choice of how to acquire a model: total survey, short-cut multiplier methods, adjusting the national model to represent the region, or purchasing a ready-made model.[11] Construction of a model entails further choices; computer software to facilitate the model construction process; sampling procedures; and regionalization procedures.

Regardless of whether a model is to be constructed or purchased, there are several tasks common to all users. First, the sectors must be defined so that they will provide the detail needed for varied uses of the model. Secondly, even users who plan to purchase a model should provide as much regional data to the modeler as possible, The importance of the collection-of-data step cannot be emphasized enough. Bourque (1987) suggests that it should include use of non-traditional data sources. Although the value of secondary, published data sources cannot be overlooked, many times information from local agencies, governments, and business can also prove valuable. Searching out individuals knowledgeable about the region, such as county extension agents or local businessmen who may have marketing data is the secret to the most accurate and robust studies.

While regional models can be constructed using several methodologies, most regional models rely on national data as proxies for at least some of the relationships expected to exist in the region. This is most critical in non-survey models, where the entire technical coefficients table for the nation is used as a proxy for the regional technology. However, modelers of the most exhaustive survey-based models acknowledge that some sectors are based on national technology. Thus, almost every regional model is a hybrid of survey and non-survey data. Therefore, the assumptions inherent in using national data should be explicitly stated. The entire national direct requirements table could be used as a regional table by assuming that the spending patterns of the firms and households in the region were the same as in the nation. But, this would be a sweeping assumption. It can be examined more closely if it is broken down into four assumptions:

(1) the product mix of the firms represented in U.S sectors is the same as that in the region;

(2) the technology to produce each given product is the same for firms in the United States and the region;

(3) the spending patterns of consumers and government are the same in the region as in the United States in general;

(4) the same proportion of each purchase is made from within the region as is made from within the United States (i.e., the same proportion of inputs is brought into the region from outside its borders as is imported to the U.S.). This requires that the region be as self-sufficient as the U.S.

An awareness of the nature of these assumptions alone will not make a regional model more accurate. However, explicit consideration of the validity of each assumption facilitates specific adjustments to more accurately depict the regional patterns.[12] In the event that data or methods do not enable such adjustments, then the analyst is responsible for interpreting results in light of these known inaccuracies and informing users of the biases and expected direction of misstatement.

If a choice is made to construct a regional input-output model based on a survey of the region, there are many steps which eliminate the need for these assumptions. If the decision is to construct or buy a model based on non-survey methodology the analyst faces a wide range of choices in procedures to eliminate the need for at least some of the assumptions of simularity of regional and national patterns. Typically, the assumption of similar technology is maintained unless expert or primary data is used to construct a region-specific production function.

After the procedures have been chosen and performed, the difficult step of drawing conclusions remains. Due to the many pitfalls of economic analysis (lack of data, sampling error, unclear definitions, and misspecified models), this step must be taken with care and with an awareness of the assumptions outlined above. Data definitions need to be clearly presented. For example, in the statement, "there are 1200 jobs in Region X," what is a job? Usually the definition will be determined not by the "desired" definition, but by the available data. Even if a variable, employment, is used that is not your ideal it is important to define exactly what it does count.

The uniqueness of individuals and of individual economic units (firms, farms, households) creates an aggregation problem. If an industry is defined as the retail sector, we know that within that sector are a myriad of jobs - probably great variation in hours worked per week, commitment to job, job skills, etc., yet we say there are 5000 retail jobs as if they are all the same. The definition and aggregation problems are more severe when models attempt to predict as well as describe. Not only are we defining all retail jobs the same, but also assuming that all retail job-holders will act in the same manner over time.

Assumptions about firms' use of resources and response to new information and technology make big differences in impacts but are difficult to determine accurately.

Use

More important than building an input-output model for the region is *how it is used*. Even the best planned and most accurate model will yield inaccurate results if it is inaccurately used. The two most common mistakes are inaccurate identification of the initial change in final demand (for which impacts are to be determined) and use of an inappropriate multiplier for the given change. At first glance, it seems efficient to make the multipliers available to all users and enable them to do their own impact analysis. However, experience has shown that this can lead to misuse of the multipliers and non-use of much additional information that the model can provide. It is better to make potential users aware of the model and its uses and encourage them to contact the modeler to do the analysis. This strategy enables the modelers and users to share their respective knowledge about the models' definitions and the sector's characteristics to correctly identify the initial change in the regional economy. This is an opportunity for the modeler to probe the user for the most detailed, appropriate information and also to explain the limitations of the model. When the study includes prescriptive actions, a specific goal must be determined to evaluate alternative strategies (i.e., relative priorities of income, jobs, and environment will impact the ranking of different development strategies).

When using the model for impact analysis or other applications, users encounter many questions too specific to be addressed in most discussions of input-output[13]. Appendix C includes a checklist to help avoid commonly-experienced pitfalls in final demand change identification and multiplier use as well as to get the most out of the model.

Summary

Input-output analysis is a useful and productive tool for regional analysis. It can provide important and timely information on the interrelationships in a regional economy and the impacts of changes on that economy. Thus, it can provide pertinent information about the impacts of economic growth and/or decline and the relative benefits and costs of alternative development strategies. Recently, the combination of a wealth of economic development issues to which input-output analysis can be applied and increased availability of computerized input-output models have led to an increased interest in this technique.

The major parts of a regional input-output model are a transaction table, a direct requirements table, and a total requirements table. From these, a variety of impact measures can be determined. Most frequently requested are multipliers. The construction and use of these tables and the associated measures are rarely straightforward and often complex. As with any analytical tool, users must be familiar with underlying assumptions, data requirements, and appropriate and inappropriate uses of it. At one time, the often prohibitive costs of constructing an input-output model for a region via the use of survey data limited its use. This changed with the development of methodologies that allow the estimation of a regional model from secondary data. The availability has continued to increase as "ready-made" models and the computer software required to estimate regional models have become available. With the construction or acquisition of a model less of a burden, more resources can and should be available for the remaining phases in conducting regional input-output analysis, planning and use. If the full benefits of this technique are to be realized, planning is required whether the model is purchased or constructed. Most importantly, the ultimate use of an input-output model and the measures associated with it require the analyst to be familiar with the inherent strengths and weaknesses of the technique.

Notes

[1] This renewed interest and use has been fueled by the decreasing cost and increased availability of the computer hardware and software required for this analysis.

[2] In this example, households are considered part of the final payments or non-processing sector; therefore, it is an example of an *open* or Type 1 model where households are not assumed to generate further new transactions in the region. If households are considered part of the processing sector and assumed to generate further transactions in a given household column pattern, then the model is a *closed* or Type II model.

[3] In practice this may require reconciliation of observed total sales and total expenditures using unallocated residual categories. For detailed suggestions concerning survey-based model reconciliation, see [Isard, Romanoff, (1971)].

[4] Theoretically, the direct requirements table includes only the production sectors. Other sectors are included in Table 1.2 for discussion purposes.

[5] It is also referred to as the technical coefficients table, alluding to the technical input ratios needed to produce the good. This title is more appropriate for a national model where many fewer of the inputs are imported. In a regional model, if the technical requirement for plastic inputs is one-fifth of the total expenditure, but plastic is not produced within the region, it will be

imported and the direct requirement (local) coefficient for the regional plastic sector will be zero.

[6] The fixed proportion assumption is based on the ratio of aggregate manufacturing output and aggregate manufacturing income and employment (121 jobs *on average* per million dollars of output), then it is assumed that new, marginal $1,000,000 will generate 121 new jobs directly. No allowance is made for the possibility (probability) that such an increase might be accomplished by an existing labor force working more productively or the increase might require additional construction workers if physical capacity has been reached. Furthermore, it is assumed that the required new workers with appropriate skills are available in the region.

[7] Miller and Blair (1985) call this the Simple Income Multiplier but this seems to be confusing with other income multipliers.

[8] It is customary to show these employment effects per $100,000 or per $1,000,000 of sales rather than one dollar because coefficients per one dollar tend to be very small values (.000298 workers per dollar of output). This is more easily understood as 29.8 workers per $100,000 of output.

[9] Assuming employment in manufacturing is 100, trade is 1000, services is 1200; then the physical labor coefficients will be Agriculture .298, Manufacturing .21 (100/467), Trade .36 (1000/2779), and Service .40 (1200/3033).

[10] Several efforts have been made to more accurately depict the additional wage spending. Referred to alternatively at Type III and Type IV income multipliers, see the following for more information, Miller and Blair, pp. 110 - 112 (1985), Palmer, Siverts & Sullivan (1985), Madden and Batey (1983), and Conway (1977).

[11] For more detail, see Brucker, Hastings and Latham (1987) and Brucker and Hastings (1986).

[12] For a detailed discussion, see Brucker and Hastings, 1986.

[13] Beemiller, Ambargis and Friedenberg handbook works through detailed examples (1986), and Bourque (1987) given many detailed hints.

Bibliography

Beemiller, Richard M., Zoe O. Ambargis, and Howard L. Friedenberg. *Regional Multipliers: A User Handbook for the Regional Input-Output Modeling System (RIMS II)*. U.S. Department of Commerce, Bureau of Economic Analysis, U.S. Government Printing Office, Washington, D.C. 20402, May 1986.

Bills, Nelson L. and Barr, Alfred L. *An Input-Output Analysis of the Upper South Branch Valley of West Virginia*. Bulletin 568T. West Virginia Agricultural Experiment Station, June 1968.

Bourque, Phillip J. *The Washington State Input-Output Study for 1982*. Seattle: University of Washington, Graduate School of Business Administration. March 1987.

Brucker, Sharon M. "New Measures from Input-Output Studies: A Comparison of Traditional Multipliers and Growth Equalized Multipliers." *Journal of the Northeastern Agricultural Economics Council*, Vol. IX, No. 2. 1980, pp. 69-74.

Brucker, Sharon M. and Steven E. Hastings. *An Interindustry Analysis of the Delaware Economy*, Agricultural Experiment Station Bulletin 452, University of Delaware. Newark, Delaware. March 1984.

Brucker, Sharon M. and Steven E. Hastings. Choices to Consider When Acquiring and Using Input-Output Models for Regional Economic Analysis. Paper presented at the Southern Agricultural Economics Association Meetings, Orlando, Florida. 1986.

Brucker, Sharon M., Steven E. Hastings and William R. Latham. "Regional Input-Output Analysis: A Comparison of Five Ready-Made Model Systems." *Review of Regional Studies*. Spring, 1987.

Burford, Roger L. and Joseph L. Katz. "A Method for Estimation of Input-Output Type Output Multipliers When No I-O Model Exists," *Journal of Regional Science*, V, 21, No. 2, May 1981, pp. 151-161.

Cartwright, Joseph U., Richard M. Beemiller and Richard D. Gustily. *RIMSII Regional Input-Output Modeling System*, U.S. Department of Commerce, Bureau of Economic Analysis, U.S. Government Printing Office, Washington, D.C. 20402. April 1981.

Conway, Richard S., Jr. "The Stability of Regional Input Output Multipliers." *Environment and Planning*, A 9, No. 2 (February 1977): 197-214.

Isard, Walter. *Methods of Regional Analysis: An Introduction to Regional Science*. New York: The Technology Press of MIT and John Wiley and Sons, Inc., 1960.

Isard, Walter and Thomas W. Langford. *Regional Input-Output Study: Recollections, Reflections and Diverse Notes on the Philadelphia Experience*. Department of Regional Sciences, University of Pennsylvania. Massachusetts: MIT Press, 1971.

Lamphear, Charles F. And Ronald T. Konecny. *ADOTMATR's User's Manual*. University of Nebraska, Lincoln, Nebraska.

Leontief, Wassily et al. *Studies in the Structure of the American Economy*. New York: Oxford University Press, 1953.

Madden, Moss, and Peter W. J. Batey. "Linked Population and Economic Models: Some Methodological Issues in Forecasting, Analysis, and Policy Optimization." *Journal of Regional Science* 23, no 2 (April 1983): 141-164.

Miernyk, William H. *The Elements of Input-Output Analysis*. New York: Random House. 1965.

Miernyk, William H. et al. *Impact of the Space Program on a Local Economy: An Input-Output Analysis*. Morgantown, W. Va. West Virginia University Library, 1967.

Miller, R.E. and P.D. Blair. *Input-Output Analysis: Foundations and Extensions*. Englewood Cliffs: Prentice-Hall, 1985.

Palmer, Charles, Eric Siverts and Joy Sullivan. "IMPLAN, Version 1.1: Analysis Guide" United States Department of Agriculture, Forest Service, Land Management Planning Systems Section (Fort Collins, Colorado: Forest Service). July 1985.

Richardson, Harry W. *Input-Output and Regional Economics*. New York: John Wiley and Sons (Halsted Press), 1972.

Schaffer, William A. "Estimating Regional Input-Output Coefficients," *Review of Regional Studies*, 2, 57-71, 1972.

Schaffer, William A. and Laurence S. Davidson. "Economic Impact of the Falcons on Atlanta: 1984." (Atlanta: The Atlanta Falcons) 1985.

Stevens, Benjamin H., George I. Treyz, David J. Erhlich and James R. Bower. "A New Technique for the Construction of Non-Survey Regional Input-Output Models and Comparisons with Two Survey-Based Models," *International Regional Science Review*, Vol. 8, No. 3, December 1983.

Stevens, B. H. and G. A. Trainer. "The Generation of Error in Regional Input-Output Impact Models," Working papers A1-76, Regional Science Institute, 1976.

Trenchi, Peter III, and Warren A. Flick. An Input-Output Model of Alabama's Economy: Understanding Forestry's Role. Bulletin 534. Alabama Agricultural Experiment Station. July 1982.

Table 1.1. Transactions Table - Region X

	Purchasing Sectors (in $ 1,000)				Final Demand		
	Ag	Manu	Trade	Serv	HH	Other	Total Output
Agriculture	202	182	10	47	100	200	741
Manufacturing	34	68	2	26	39	298	467
Trade	47	35	991	440	1200	66	2779
Service	86	59	565	510	1500	313	3033
Households	200	40	205	1250	200	1494	3389
Imports	172	83	1006	760	350	1053	3424
Total	741	467	2779	3033	3389	3424	

Table 1.2 Direct Requirements Table - Region X

	Purchasing Sectors				
	Ag.	Manu.	Trade	Service	Households
Agriculture	.27	.39	.00	.02	.03
Manufacturing	.05	.15	.00	.01	.01
Trade	.06	.07	.36	.15	.35
Service	.12	.13	.20	.17	.44
Total local purchases	.50	.73	.56	.35	.83
Households	.27	.08	.07	.41	.07
Imports	.23	.18	.36	.24	.10
Total	1.00	1.00	1.00	1.00	1.00

Table 1.3 Total Requirements Table - Region X
(Direct and Indirect Requirements)

	PURCHASING SECTORS			
	Ag.	Manu.	Trade	Service
Agriculture	1.42	.66	.01	.04
Manufacturing	.09	1.22	.01	.02
Trade	.21	.28	1.66	.31
Service	.27	.35	.40	1.29
Total	1.99	2.51	2.08	1.66

Table 1.4 Total Requirements Table - Region X
(Direct and Indirect and Induced)

	PURCHASING SECTORS			
	Ag.	Manu.	Trade	Service
Agriculture	1.49	.72	.05	.12
Manufacturing	.11	1.24	.02	.04
Trade	.87	.85	2.03	1.03
Service	.93	.92	.77	2.01
Total	4.76	4.88	3.37	4.21

Table 1.5 Total Income Effects

	Ag.	Manu.	Trade	Service
Agriculture	0.402	0.194	0.014	0.032
Manufacturing	0.009	0.099	0.002	0.003
Trade	0.061	0.060	0.142	0.072
Service	0.381	0.377	0.316	0.824
Household	0.064	0.055	0.036	0.070
Total	0.918	0.785	0.509	1.002

Table 1.6 Total Employment Effects (per $100,000)

	Ag.	Manu.	Trade	Service
Agriculture	44.4	21.5	1.5	3.6
Manufacturing	2.4	26.6	0.4	0.9
Trade		31.3	30.6	7330.1
Service	36.8	36.4	30.5	79.5
Household	8.1	6.9	4.5	8.9
Total	123.0	121.9	109.9	129.9

Appendix A

Algebraic Formulation of a Simple Input-Output Model

Following the notation used by Miller and Blair (1985), the basic structure of input-output analysis can be presented. All economic activity in a region is divided into n production sectors. The total output of each sector is determined by the intermediate demand (purchases by other production sectors) and final demand (purchases by households, government, exports, etc.) for the sector's products. In a two sector economy,

$$(1.0)\quad X_1 = z_{11} + z_{12} + Y_1$$
$$(1.1)\quad X_2 = z_{21} + z_{22} + Y_2$$

where, X^1 = the total output of sector 1,

z_{11} = the flow of products from sector 1 to sector 1,

Y_1 = the final demand for the production of sector 1.

The flow of products, z, from one sector to another is determined by the total output of the purchasing sector. Furthermore, each z is a fixed and constant proportion of the total output of the purchasing sector. For each sector, this proportion (a) is

$$(2.0)\ a_{11} = z_{11} / X_1$$
$$(2.1)\ a_{12} = z_{12} / X_2$$
$$(2.2)\ a_{21} = z_{22} / X_1$$
$$(2.3)\ a_{22} = z_{22} / X_2$$

Solving Equations (2.0 - 2.3) for z's (i.e., $z_{11} = a_{11} * X_1$) and substituting into Equations (1.0 - 1.1), then Equations (1.0 - 1.1)) become

$$(3.0)\ X_1 = a_{11} X_1 + a_{12} X_2 + Y_1$$
$$(3.1)\ X_2 = a_{21} X_1 + a_{22} X_2 + Y_2$$

The values of X_1 and X_2 are considered endogenous and the values of Y_1 and Y_2 are considered exogenous. Therefore, the level of final demand for the various sectors output determine the total output of the production sectors.

A similar equation exists for each sector in the economy, therefore, this system of equations can be represented in matrix notation as

$$(4)\ X = AX + Y.$$

where, $X = \begin{bmatrix} X_1 \\ X_2 \end{bmatrix}$ $A = \begin{bmatrix} a_{11} & a_{12} \\ a_{21} & a_{22} \end{bmatrix}$ $Y = \begin{bmatrix} Y_1 \\ Y_2 \end{bmatrix}$

Solving Equation (4) for X, the outputs of the production sectors equal

$$(5)\ (I - A)X = Y$$

$$(6)\ X = (I - A)^{-1}Y$$

where, I = a 2 x 2 identity matrix.

Using this relationship, the outputs of the production sectors can be determined for any projected levels of final demand. It is assumed that the relationships identified initially (Equation (4)) hold in determining the new levels of final demand and output.

Appendix B

Table 1A: Output and Income Measures from Input-Output Analysis, Region X

Sector Name	Output Multiplier		Total Income Effects		Income Effects			Income Multipliers	
Column	Type I[1] (1)	Type II[2] (2)	Open[3] (3)	Closed[4] (4)	Direct[5] (5)	Indirect[6] (6)	Induced[7] (7)	Type I[8] (8)	Type II[9] (9)
Agriculture	1.99	4.76	0.52	0.92	0.27	0.25	0.40	1.92	3.41
Manufacturing	2.51	4.88	0.44	0.78	0.08	0.36	0.34	5.50	9.75
Trade	2.08	3.37	0.28	0.51	0.07	0.21	0.23	4.00	7.29
Service	1.65	4.21	0.56	1.00	0.41	0.15	0.44	1.36	2.44

[1] Total row from open model (Table 3).
[2] Total row from closed model (Table 4).
[3] Total (direct and indirect) income effects per $1 of final demand. TO CALCULATE: Multiply each element of the sector's total requirements column (Table 3) by the direct income effect (column 5) and sum.
[4] Total (direct, indirect and induced) income effects per $1 of final demand. TO CALCULATE: Multiply each element of the sector's total requirements column (Table 4) by the direct income effect (column 5) and sum.
[5] Household row of the direct requirements table (Table 2).
[6] Total income effects from open model (column 3) minus the direct effect (column 5).
[7] Total income effects from the closed model (column 4) minus total income effects from the open model (column 3).
[8] Total (direct and indirect) income effects per $1 change in initial income. TO CALCULATE: Divide the total income effect from an open model (column 3) by the direct income effect (column 5).
[9] Total (direct, indirect and induced) income effects per $1 change in initial income. TO CALCULATE: Divide the total income effect from a closed model (column 4) by the direct income effect (column 5).

Appendix B

Table 1B: Employment Measures from Input-Output Analysis, Region X

Sector Name	Sector Employment[10]	Total Employment Effects		Employment Effects			Employment Multipliers	
		Open[11]	Closed[12]	Direct[13]	Indirect[14]	Induced[15]	Type I[16]	Type II[17]
			(Jobs per $100,000 final demand)					
Column	(10)	(11)	(12)	(13)	(14)	(15)	(16)	(17)
Agriculture	221	63	123	30	33	60	2.09	4.10
Manufacturing	100	69	122	21	49	53	3.31	5.81
Trade	1000	76	110	36	40	34	2.11	3.06
Service	1200	64	130	40	24	66	1.61	3.25

[10] Assumed employment for Region X; employment data would be provided by the analyst.

[11] Total (direct and indirect) employment effects per $100,000 of final demand. TO CALCULATE: Multiply each element of the sector's total requirements column (Table 4) by the direct employment effect (column 13) and then sum.

[12] Total (direct, indirect and induced) employment effects per $100,000 of final demand. TO CALCULATE: Multiply each of the sector's total requirements (column 2) by the direct employment effect (column 13) and then sum.

[13] Total sector employment (column 10) divided by sector output in $100,000 (Table 1).

[14] Total employment effects from open model (column 11) minus the direct employment effects (column 13).

[15] Total employment effect from closed model (column 12) minus the total employment effect of the open model (column 11).

[16] Total (direct and indirect) employment effects per change in initial employment. TO CALCULATE: Divide the total employment effect from the open model (column 11) by the direct employment effect (column 13).

[17] The total (direct, indirect and induced) employment effects per change in initial employment. TO CALCULATE: Divide the total employment effect from a closed model (column 12) by the direct employment effect (column 13).

Appendix C
Input-Output Checklist

Identification of Initial Change in Final Demand

1. What Is the Value and Sector of Change?

Verify that it is really Final Demand (exports, consumers, or government) and not production for other processing sectors. Also, make sure that the dollar values given are consistent with the measure of output defined for the sector. For example, when working with retail trade, output is typically based on value added (total sales minus merchandise purchased for resale); while the dollar value that a user supplies is likely to be based on total sales. Adjustment is necessary because total sales would overstate the amount of retail output change. This same problem can occur with banking, transportation, or real estate, depending on sector definitions.

2. Are Data on Employment or Income Changes Available?

Allow users to define their initial change as an increase or decrease in employment or income in the region when this is the only data they have available. Then use the income and employment multipliers to determine region-wide impacts.

3. Are There Offsetting Changes in the Economy?

When a new company opens up a factory, it is assumed that the model can use multipliers to estimate the impact. Usually this is true, but it is technically important to make sure that the firm is not simply taking the place of another industry in the region which would be decreasing its output; the two impacts would probably offset one another. Conversely, if one firm is decreasing production, check what other firms in the sector are doing. They may be picking up slack resources and increasing their output, offsetting the initial change.

4. Is New Construction Required by the Change?

If a new facility is to be built, the impact of the construction should also be estimated. Recalling from above, this is a separate impact because input-output does not include capital account purchases.

5. Does Final Demand Change in More Than One Sector?

If the decline or growth is not in a specific industry but rather an increase in overall income in the region, remember that the assumption is new final demand. Therefore, if there is increased income, you have to identify which sectors experience the increase in final demand. If the nation as a whole is growing, then the final demand increase is for some basket of goods from your region. Identify these and then determine the impact of each sector's growth or decline of final demand, and sum the impacts. This method can also be used for a change (like population growth) that impacts several sectors at once.

Use of Multipliers

1. Which Multiplier Is Appropriate?

The appropriate multiplier to use depends both on the information desired and the information that can be provided. Appendix B provides a summary of these measures for hypothetical Region X.

Output Multipliers (Type I): (Column 1, Appendix B) Use when change in final demand is known and total (direct and indirect) change in regionwide production (output) is desired.

Output Multipliers (Type II): (Column 2, Appendix B) Use when change in final demand is known and total (direct, indirect and induced) change in regionwide production (output) is desired.

Total Income Effects (Open or Closed Model): (Column 3 and 4, Appendix B) Use when change in final demand is known and total (for direct and indirect, use Open; for direct, indirect and induced, use Closed) change in regionwide income is desired. To use this measure for disaggregated impacts, multiply the increase or decrease in final demand by the income effects column for that sector in the total income effects table. (SHORT CUT: If only the total effects are needed, the appropriate coefficient from the household row of the total requirements table can be multiplied by the change in final demand.)

Income Multipliers (Type I): (Column 8, Appendix B): Use when the initial change in sector income is known and the total (direct and indirect) change in regionwide income (all sectors combined) is desired.

Income Multipliers (Type II): (Column 9, Appendix B): Use when the initial change in sector income is known and the total (direct, indirect and induced) change in regionwide income (all sectors combined) is desired.

Total Employment Effects (Open or Closed): (Columns 11 and 12, Appendix B) Use when change in final demand is known and total (for direct and indirect, use Open; for direct, indirect and induced, use Closed) change in regionwide employment is desired. Effects in Appendix B refer to change in jobs associated with a $100,000 change in final demand for a sector.

Employment Multipliers (Type I): (Column 16, Appendix B) Use when the change in the number of employees who will be employed or laid off is known and the total (direct and induced) change in regionwide employment is desired.

Employment Multipliers (Type II): (Column 17, Appendix B) Use when the change in the number of employees who will be employed or laid off is known and the total (direct, indirect and induced) change in regionwide employment is desired.

2. What Do Employment Multipliers Imply?

Employment multipliers mean that the initial change will result in the multiplied number of jobs if the ratio of total employees to production in each sector is the same for additional production as in the ratio for that sector in the original model, and if the unemployed people in the region have the skills required on the new jobs.

3. Are Big Multipliers Better?

If size of a sector's multiplier is being used to evaluate targets for growth, the planner should be introduced to some measure of the feasibility of certain growth patterns, such as Growth Equalized Multipliers, or elasticities (Brucker, 1980) or a simple comparison of the region-wide impact resulting from standard percentage growth of several sectors. A sector with the largest multiplier in the state may be so small that it takes an unrealistic rate of growth to generate the same region-wide growth of income as a very large sector with a very small multiplier.

Getting the Most from the Model

1. Are Multipliers Enough?

Although I-O is best known for measuring region-wide effects, this only scratches the surface of its potential uses. The power of the model is that it can also show the distribution of the overall impacts. A column of the Total Requirements Table indicates which sectors in the region will be affected and by what magnitude. When translated into income and employment effects, this can be used to make important policy decisions. Although the model assumes an unlimited supply of all resources, the user's knowledge of the region may identify a sector which does not presently have the capacity to meet the increased demand predicted. Policy makers can use this information to identify an industrial growth target for the region or to target unemployment and youth job training programs.

2. Are You Ready for Inquiries?

Perhaps the most universal experience of I-O modelers is that the demand for information from the model will exceed your ability to provide it; be prepared to deal with this. Have forms prepared to formalize the identification of changes you are to analyze, have the model available on a personal computer with software that facilitates utilizing the multipliers and coefficients for both single and multiple demand change scenarios.

3. Do You Improve It by Interacting with Users/Clients?

Researchers should capitalize on the knowledge that users/clients may have about industry's expenditure patterns, the nature of the impact, and/or overall economic conditions in the region to improve the accuracy of the estimated model and assorted impacts.

2

Cautions in Using I-O Models

Mark S. Henry and Thomas G. Johnson

Introduction

The use and abuse of Input/Output (I-O) models for analyzing change in regional economies has a history spanning some three decades. The increased availability of microcomputer versions of I-O models over the past several years holds promise for more effective use of the technique. However, effectiveness is sometimes thought of as quick turnaround, low cost, easy to read tabular results, and pleasing graphics. Certainly, the ease with which I-O results can be made available and interpreted are important to users. Our purpose is to emphasize the need to make I-O analysis effective by proper structuring of the regional problem within the I-O framework, appropriate uses and interpretations of the model and by acknowledgement of the limitations of I-O. In short, what are the major "do's and don'ts" once you have an I-O model for your region.

Accounting Conventions and Reduction Procedures

The availability of software like IMPLAN and ADOTMATR[1] is likely to vastly improve the quality of regional I-O models for regional analysis. ADOTMATR provides step-by-step instructions in their user manual, the national Use and Make matrices at the 528 sector level, and LOTUS-like commands for helping the analyst to build and use regional I-O models. IMPLAN also provides the data needed for a basic I-O framework and Social Accounting Matrices (SAM's) from the national to the county level. It also employs a user-friendly series of instruction menus that both create the I-O model and carry out analyses. Both of these systems -- as well as most of their competitors -- may provide more consistent procedures across regions in building nonsurvey I-O models. This represents an important step in reducing one problem with I-O modelling across regions -- differences in accounting procedures. For example, with IMPLAN, the models are constructed so that there is data consistency from the "top-down," i.e., from nation to state to county. After applying regional purchase coefficients (RPC's), each model is balanced to meet the required I-O accounting definitions.

Analysts using ADOTMATR may use a variety of nonsurvey reduction procedures to reduce the national models to reflect regional industry mix and import requirements. Since the jury is still out on which nonsurvey model or method of creating a regional model is most accurate and is likely to remain so, users of I-O models should recognize that answers to their questions about the impact of the new plant, etc. will still vary with the I-O model used. The message for the multi-region analyst is to use the same I-O model procedures across regions. This means using the same reduction and accounting procedures as the region of interest changes. Users may decide to use one of the hybrid models like GRIMP from Australia. While these procedures allow entry of primary or superior data into the non-survey framework when available, the message is the same. Use the same accounting and reduction procedures across regions.

Defining the Region

The region of interest for the I-O model user will often be some political subdivision. Users should recognize that their political unit -- a county or multicounty planning area -- is part of the broader regional economy that is likely to be affected by a much wider array of forces than those that directly affect its member subregions. Ideally, interregional models that capture all of the import and export flows between subregions are needed to understand how the subregion of interest will be affected by some set of exogenous forces. This kind of model is rarely available so definition of the region of interest becomes quite important. Although county level models may be acceptable in some cases, a general guideline is to use a multicounty unit as the minimum geographical size with allocations to counties or other smaller units as the data permit. This reduces problems of evaluating such things as employment impacts and their associated consumption impacts over a commuting area from a change like a new plant in one community. Indeed, it is often desirable to define an impact region based on the expected commuter fields to the new plant. For impacts associated with other exogenous events like tourism, the geographical scope will largely be determined by the area designated as the destination of tourists and its commuting field.

Generally, the smaller the region, the more important export and imports become and the coefficients that reflect these flows may be quite sensitive to forces of economic change. All else the same, larger regions tend to have fewer leakages of savings, taxes and imports so that multiplier effects are larger for larger areas. This simple idea needs to be kept in mind when modelling the impact of an event that is likely to have multicounty impacts. The sum of the impacts from say three single county models is likely to be less than the impacts estimated for the same event from a three county model.

Some Suggestions After the I-O Model Is Built

After consideration is given to the appropriate accounting and reduction procedures as well as the geographical dimension of the model, some important additional issues involve proper use and interpretation of the I-O model. We next discuss three areas of I-O user education: (1) structuring the I-O problem; (2) interpreting I-O multipliers, and (3) some commonly made misinterpretations of the I-O framework. Some guidelines for users that will enable them to avoid the most serious problems of misapplication and misinterpretation of their I-O models are suggested in a final section.

Structuring the I-O Problem

Probably the best suggestion to make here is to keep in mind the double entry accounting basis used to represent the flows of the economic system that are depicted with the I-O model. Total sales across the row representing any sector are decomposed as intermediate and final sales. These total sales in the industry by industry accounts in turn must equal the sum of the value of intermediate and final purchases required by that sector. For most purposes, recalling that gross output for any given sector is identical to the sum of sales to intermediate users and final users (final demand) is a good starting point for structuring the I-O problem. In the most basic industry by industry system of accounts:

$$(1) \quad X = AX + Y$$

where

$X =$ the total gross output vector

$A =$ the matrix of direct input coefficients and,

$Y =$ the vector of final demands aggregated over consumption, investment, government spending and net exports

In the usual case, the vector of final demands is exogenous with the X vector the unknown to be determined by solving equation (1) for X :

$$(2) \quad X = (I-A)^{-1} * Y$$

This solution requires the following structuring of the I-O problem: First, the exogenous event must be expressed as some change in Y the final demand vector. This is a seemingly easy task yet in practice can be quite demanding of resources and creativity. Recall that the final demand vector is composed of personal consumption expenditures by sector, investment expenditures by sector, government purchases by sector, and net exports by sector. Even simple multicounty models are likely to have several hundred sectors active in the region if one is working at the most disaggregated level possible (highly recommended by all I-O analysts).[2] And, the remaining sectors (there are 528 interindustry sectors in the national model) are also important to identify as possible sources of import substitution in the region.

Consider the steps one might take to create the proper final demand accounts to undertake a basic impact analysis of an exogenous event (stimulated by outside government spending) -- say a new nuclear production reactor for producing tritium at the Savannah River Plant near Aiken, South Carolina. When an open model (the preferred form for most analyses) is used, the consumption column and household row are not included in the A matrix. In this case, three issues arise. First, if the construction impacts are to be examined, a vector of purchases of construction materials by economic sector and final payments to labor, management, etc., is constructed (see Figure 1). Each of these entries must be translated into the prices of the base year of the I-O model (see Figure 2). Next, each of the entries should be adjusted to reflect local capacities to supply these materials -- i.e., even if the supplying sector exists in the region, can it be expected to be able to supply all or only part of the local needs? If the sector does not currently exist in the region, is import substitution likely over the term of the project? If so, a new model incorporating the new sector at some future point will be needed -- that is the economic restructuring problems should be addressed.

The payments to labor, management, etc., during the construction phase will generate personal consumption expenditures (PCE). These PCE expenditures will need to be identified by type of worker -- in-migrants, commuters, and current residents. Each is likely to have somewhat different local PCE expenditure patterns. These PCE vectors also have to be translated into the appropriate sectors with careful attention to trade margins when assigning PCE's to industry sectors (local and imported) and the various margin accounts -- mainly trade and transportation.

Similar procedures are followed to build PCE vectors for the operating phase of the new plant. In the open version of the model, the destination of the sales of the plant need to be recognized. In most cases, it is likely that the destination of the plant's sales can be assigned (with the appropriate margin adjustments) to final demand components: PCE, Investment, Government purchases, or Net exports. However, it is also quite likely that some of the plant's sales are sold to other sectors as an intermediate purchase. In the I-O accounts, the impact of this intermediate sale should appear as a final demand sale by the purchasing sector that is associated with this intermediate transaction. For example, exports of manufactured tobacco products require intermediate purchases from the tobacco stemming and redrying sector, which in turn purchases raw tobacco from the farm level tobacco sector (see Figure 3).

For the SRP case however, it is not clear where the final product, tritium, goes and it is somewhat awkward to think of an NPR sector. Here it is best to construct a vector of intermediate purchases made by the NPR -- much like during the construction phase -- to identify local industries that supply inputs directly to the NPR.

Miller and Blair (1985) describe many variations on the basic model depicted in equations 1 and 2. Some of the sectors may be treated as exogenous and some endogenous. Once the mixed models are adopted however, a new set of equations must be solved and it is not clear that the ADOTMATR's and IMPLAN's are well equipped to solve these problems.

In sum, most exogenous economic events can be translated into a final demand format. This may be as simple as increased exports of a product from an existing sector in the region -- exports of cigarettes from the state of Virginia. Here the only problem is to construct the final demand change to reflect producers value of the cigarettes in Virginia, and to estimate transportation, insurance and trade margins associated with the FAS export value. If one is working with the open I-O model, then induced impacts will have to be estimated separately as discussed below. Major construction projects with subsequent operating impacts require a good deal of effort to construct the final demand accounts needed to make use of the Leontief inverse for estimation of total impacts. Tourism, recreational use of national forests, wholesaling, service based activities, and other activities that are servicing final demand can all be put into the I-O framework to evaluate their impacts on the region. But often the process of structuring the I-O problem will require careful attention to the I-O accounting conventions and use of primary data to create the final demand vectors that reflect the nature of the exogenous event.

Interpretation of I-O Multipliers

The nice feature of using the final demand approach to doing I-O analysis is the ability to estimate sector by sector impacts for any exogenous event. That is, with the appropriate set of conversion ratios, total gross output, income and employment impacts by sector can be computed. These may include both direct, indirect and household induced effects by sector from the exogenous event. When available, the sectoral detail in an I-O model is preferred to using the next best alternative, multipliers. Multipliers are less desirable than the final demand approach because one cannot sort out the indirect and induced impacts by economic sector. This information may be critical for those who are interested in identifying potential import substitutes as part of a regional economic development strategy. If one is forced to use multipliers, the potential -- unintended or otherwise -- for abuse also tends to rise. A good way to reduce the inappropriate uses of I-O multipliers is to follow the guidelines suggested by the authors of RIMS (see Figure 4). Using the open model versions, gross output, earnings and employment multipliers are listed for each sector in South Carolina. It is important to note that these are all final demand driven multipliers. The first two multipliers translate a one dollar *delivery to final demand by the sector* into either gross output, or earnings in all sectors from that one dollar delivery to final demand. The employment multiplier is an estimate of the number of jobs generated for a $1 million delivery to final demand.

Note that these multipliers do not incorporate the household income/consumption, induced multiplier effects from the delivery to final demand. The income that is earned as a result of the final demand delivery and the associated indirect output requirements and respent on consumer goods in the region may be captured in two basic ways. First, the closed version of the I-O model which includes the household row and column in the A matrix captures all of the income/consumption expenditure effects. Gross output multipliers driven by final demand changes can be calculated from the closed Leontief inverse by summing down the rows for the column sector of interest -- with the usual constraint of not including the row value in the household row. This avoids mixing output and income concepts in the gross output multiplier. Earnings multipliers that include the household induced impacts can be read simply as the household row entry in the closed Leontief. Employment multipliers are calculated by multiplying the elements in a column of the closed inverse by corresponding elements of a vector of employment/gross output ratios. The sum of the element by element products -- again the household row entry is ignored -- is the final demand driven employment multiplier.

The second way of capturing the household induced impacts is to work with the open Leontief inverse and a series of income/consumption expenditure rounds induced by the initial final demand sale. The initial final demand delivery can be used to estimate direct and indirect gross outputs by sector. These gross outputs can be used to estimate household income. In turn, these income flows are translated into an incremental consumption vector using the expenditure patterns from the household column in the I-O model. Subsequent rounds of income and expenditure flows are captured by repeating this sequence until tax, import and savings leakage effects result in negligible added consumption.

Whether one chooses to close or not to close an I-O model with respect to households, depends partly on one's assumptions about length of the observation period. During short intervals after an economic shock, indirect effects are likely to dominate the economic impact because the income-expenditure cycle has had insufficient time to be affected. After longer periods, the indirect effects will be joined by the induced effects making the closure of the model assumption more appropriate. Two precautions are suggested. First, neither the open or closed based multipliers have any temporal dimension that is readily observable. So the suggestion to include induced impacts only in the longer run view is just that -- a suggestion. Second, closing the model with respect to households further strains the assumption of linearity in the I-O framework since the household sector is less likely than other sectors to change linearly with income.

IMPLAN uses the open model with induced impacts flowing from added employment to added population to added consumption on a per capita consumption basis. Long ago, Tiebout argued for the use of marginal consumption coefficient for current residents who experience increments to income from some exogenous event. In-migrants on the other hand could be treated with average coefficients since they presumably move to the area with the same average expenditure patterns as the current residents. Obviously, each of these approaches suggests a research agenda more than defining the best approach for a given assessment of induced impacts.

Note that the I-O system of accounts can be solved for the equilibrium levels of gross output as in equation 2 from either the open or closed versions of this fundamental equation by moving the household row and column in or out of the A matrix with the appropriate adjustment to the final demand vector, Y. However, once this initial equilibrium Y vector is disturbed -- say by added export activity, the new level of induced consumption in the open version of the model must be determined as a sequence of subsequent income and consumption expenditures. The importance of carefully specifying the role of household income and expenditures at the regional level is likely to be far more important to model accuracy than misspecifying the linkages between industries or the limitations of the proportionality assumption of the implied production functions.

Although the Type I, Type 2, Type 3, etc. multipliers are discussed extensively in the I-O literature, we defer to the discussion of this multiplier by Miller and Blair (1985). We prefer to stress the notion of changes in final demand as the driving force behind the various multiplier effects. Our purpose is to maintain -- in the mind of users of I-O models -- the linkage between the I-O system of accounts and the multipliers that are ultimately derived from them. If one is forced to use the various ratios of total to direct changes, it is our belief that the chances for abusing the interpretation of the I-O multiplier and the system of accounts from which they are derived. We discuss some of these issues in the next section.

Some Common Abuses

The problem of matching multipliers to their appropriate exogenous forces ("drivers") is a potentially serious abuse of I-O models by users. Analysts are frequently asked to give local officials or extension agents a multiplier for evaluating the impact of agriculture on a county economy. A typical response is usually to ask what kind of impacts are of interest to them -- gross output (sales), income, or employment? A common reply is, "which is the largest?" At this juncture, we know trouble lies ahead and that we need to think about developing some training programs for our local officials and agents. It may seem hard to believe but the myth of the "7 to 1" farm income multiplier (even at the county level) is alive and well in the minds of many agriculturalists. The following review provides insights to this enduring myth (Schluter 1983).

> In 1945 Carl Wilkins of the Raw Materials National Council of Sioux City, Iowa published a pamphlet, "A Prosperous Post-War Era is Possible," in which he showed a 1 to 7 average ratio between farm cash receipts and nominal national income. Although subsequent academic and professional reviews were extremely critical of Wilkins' 1-7 multiplier, various farm groups have periodically made reference to this multiplier... As a means of highlighting the weak theoretical base of this multiplier, we update the Wilkins analysis for subsequent 20 year time periods.
>
> An examination of this table highlights at least two problems with Wilkins analysis. His 7-1 ratio has not remained stable and his direction of causation is questionable. Does farm income introduce income into the economy to generate multiple rounds of indirect national income as implied by Wilkins or does aggregate national income provide a market, albeit an inelastic one, for farm output?
>
> Wilkins supporters need to ask: If sales of farm commodities are such powerful inducers of income flows in the economy why has the farm sectors' share of these national income flows declined over time?
>
> The answer unfortunately for income flows in the farm sector is that given the relatively inelastic price and income demand for farm commodities and the inelastic supply of these commodities, as an economy develops the farm sector can expect to decline in relative importance. With this decline in relative importance, the inverse of

the farm sector's relative importance interpreted as a multiplier shows a larger multiplier effect but this is only an accounting relationship.

This is an example of a multiplier that was clearly unrelated to the actual interdependencies that exist between farm income and total income. It provides a useful context within which to examine why I-O multipliers are not simple accounting ratios but rather reflect a mix of production technologies extant at the time the I-O model data were obtained. It is ironic that the great strength of the I-O multipliers is that they are derived from the double entry accounting system that defines the transactions (or Use and Make) flows in a economic system. Yet, they are not simple accounting identities like the Wilkins 1-7 farm income multiplier. Rather, they incorporate two fundamental behavioral assumptions.

First, the fixed proportions production function explicit in the direct requirements I-O table is an assumption about the profit maximizing behavior in each business sector. Indeed, the direct requirements table that is derived from the accounting data of the transactions (or Use and Make) flows is now properly considered to be a model of the technology in the regional economy. It is interesting to note that El-Hodiri recently demonstrated that the A_{ij} coefficients of the Leontief model in value terms are equivalent to a Cobb-Douglas production technology. Cobb-Douglas allows physical input substitution as relative prices change while input coefficients in value terms remain constant as required in Leontief's I-O system (El-Hodiri, 1988).

Second, the I-O multipliers derived from the Leontief inverse matrix rest on the Keynesian notion that exogenous change will have multiple impacts on the economy and multiplier effects will increase as leakages of taxes, savings, and imports are reduced. The key difference between the Keynesian multiplier propagation process and that in the Leontief model is the gross output (final and intermediate) basis for the Leontief variety while the Keynesian multiplier is defined in terms of final demand (final expenditures -- C, I, G, and X-M) or alternatively value added (factor incomes -- wages, rent, profits, and interest). Importantly, in the standard I-O system, it is the change in the final demand components that are the exogenous forces driving the gross output based multiplier of the Leontief system.

Summary

Abuses of I-O multipliers frequently arise from the following errors:

1. Using estimates of gross outputs of a new activity as the change in final demand. In fact much of the new sales may go to within region firms which enables them to make added sales. This error uses the wrong value

for the exogenous event. Even if the multiplier is correct, the overall impact estimate will be wrong.

2. Failing to adjust final demand sales for margin activities. For example, consumer purchases of new autos in the region should be decomposed into the value of the trade activity purchased by the consumer, the value of the transportation charge in getting the product to the consumer, and the value of the auto at the plant. Moreover, the geographical location of the margin and manufacturing activities are of great concern with regional models. Again, failure to make the proper margin allocations and sector assignments will result in serious errors even if the I-O multipliers are valid.

3. Failure to maintain the final demand vector changes in constant dollars that are pegged to the year in which the I-O model was constructed. Estimates of current dollar impacts can be made -- imprecisely -- using aggregate price indices to convert the constant dollar gross output impacts into current dollars. Another reason to avoid multipliers and use the I-O model is the ability to price adjust estimates on a sector by sector basis.

4. The Gross output multiplier in sector i is driven by final demand changes in sector i. The result of this multiplication does *not* yield an estimate of gross output *in* sector i. It does yield an estimate of the gross output changes in all sectors from the change in final demand for sector i. But individual sector impacts are concealed in the multiplier.

5. Use of multipliers from closed I-O models is likely to over-estimate the extent of income/consumer expenditure induced outputs for several reasons. I-O modelers typically close a model with respect to household by balancing household income on an estimate of total local PCE including PCE from imports. As Rose and Stevens (1988) note, typically all employee compensation in each sector is assumed to be spent and the various property type incomes by sector are used in a proportional manner to augment employee compensation if needed. This procedure ignores both outside sources of local PCE (commuters, travelers, etc.) and outside sources of income used for local PCE (e.g., transfers). So in addition to the cross hauling source of overestimation of output multipliers, Rose and Stevens note a problem with income cross-hauling in the closed I-O models providing a source of overestimation of closed I-O multipliers.

6. Use of final demand changes to drive multipliers that are based on ratios of total change to direct change. Careful attention to the tables from Miller and Blair in the Appendix avoids this problem.

7. Use of multipliers based on 1977 technology to estimate impacts 10 to 20 years later -- a problem most all I-O analysts have to live with unless they have unusually well endowed and academically curious clients.

8. Aggregation of a base year I-O model and use of the multipliers or Leontief inverse for impact analysis. Large changes in sectoral multipliers -- for sectors not aggregated -- can result. See Miller and Blair for procedures to follow to mitigate these problems.

Guidelines for Proper Use of Regional I-O Models

1. Multiplier effects should be expressed per dollar of final demand sales whenever possible (RIMS is an excellent example of this procedure) and driven by exogenous final demand changes wherever possible.

2. When gross output (sales) changes are the known exogenous change, convert standard Leontief inverses so that they express the total changes in terms of change in gross output rather than final demand sales. Or more simply, estimate the associated final demand flows.

3. Be as careful as possible in formulating a final demand vector for some exogenous change so that margin activities and origin of the manufactured product being produced for final demand delivery are carefully delineated. Effects of errors here might swamp errors in constructing regional coefficients from national technical coefficients.

4. Open Model multipliers should be used with careful consideration given to alternatives for incorporating the household induced effects through changes in the household consumption vector(s) depending on the kind of exogenous event under investigation. Use of the open model also avoids confusion with Type 2 multipliers -- truncated versus nontruncated varieties and forces the analyst to devote more effort to defining the impact of the household activities.

5. When the exogenous event is expected to have a significant impact on the structure of the local economy, make sure the appropriate system of equations is used and new multipliers calculated if needed.

6. The time dimension within which the total impacts are assumed to be felt should be recognized.

7. For small region models-recognize the importance that cross hauling may have, the problem of import coefficient instability, and the need to use local data to check on the use of data from large data bases.

8. Small regions should be functional economic areas where possible. This multicounty grouping will reduce -- though not eliminate -- the problems of cross-hauling and import coefficient instability.

Finally, perhaps the biggest abuse is insufficient use of the I-O model to analyze a wide range of natural resources and regional economic issues that have important distributional consequences. Often, public and political leaders are concerned more with who benefits and who pays for alternative policy actions rather than the relative efficiency of alternative actions. Rose, et al., (1988) have demonstrated the proper use of the I-O framework for analysis of distributional issues.

The potential for effective use of I-O at the regional level is greatly enhanced by three areas of current research: (1) the improved software for constructing models; (2) Improved data bases though local level primary data are still lacking;, and (3) By careful attention to I-O problem structuring like those by Rose et al. With improved ease of access to software and I-O data comes a rise in abuse of the I-O model. Since I-O models are derived from a double entry accounting system, it is fitting that we conclude with a plea for a set of I-O generally-accepted-accounting principles (IOGAAP). These would incorporate many of the suggestions in this volume and move us closer to an understanding of how regional economies work and respond to outside forces.

Notes

[1] IMPLAN is an I-O software package and model developed by the U.S. Forest Service, Ft. Collins and ADOTMATR and I-O software package developed by Ron Konecny, University of Nebraska.

[2] The problem of aggregation of I-O models is discussed in Miller and Blair. Changing the number of sectors in the model can have substantial effects on multiplier estimates. Miller and Blair provide some guidelines to mitigate this problem.

References

Alward, G. S., H. C. Davis, K. A. Despotakis, and E. M. Lofting. 1985. "Regional Non-Survey Input-Output analysis with IMPLAN," paper presented at the Southern Regional Science Association Conference, Washington, D. C., May 9-10, 1985.

El-Hodiri, M. and F. Nourzad. 1988. "A Note on Leontief Technology and Input Substitution." *Journal of Regional Science.* 28:1, February.

Johnson, Thomas G. 1984. "Towards More Standardized Input-Output Multipliers," unpublished paper SP-83-15, Dept. Of Agricultural Economics, VPI&SU, Blacksburg, Virginia.

Lamphear, F. C. and R. Konecny. 1986. *ADOTMATR.* Resource Economics and Management Analysis, Inc., Lincoln, Nebraska. 1986.

Miller, Ron and Peter Blair. 1985. *Input-Output Analysis: Foundations and Extensions.* Englewood Cliffs, New Jersey, Prentice Hall, Inc.

Rose, A. and Benjamin Stevens. 1988. "Transboundary Income Flows in Regional Input-Output Models: Or How to Close A Regional I-O Model Properly." Paper presented at the International Conference on Input-Output Modelling. Morgantown, West Virginia. August 13-16, 1988.

Rose, A., Brandt Stevens, and G. Davis. 1988. *Natural Resource Policy and Income Distribution.* Baltimore: John Hopkins University Press.

Schluter, G. and W. Edmonson. 1983. "Carl Wilkins and the 7-1 Farm Income Multiplier of Farm Area Folklore." Unpublished communication. Economic Research Service, U. S. Dept. of Agriculture.

Tiebout, C. M. 1969. "An Empirical Regional Input-Output Projection Model: The State of Washington 1980." *Review of Economics and Statistics.*

U.S. Department of Commerce. Bureau of Economic Analysis. 1986. *Regional Multipliers: A User Handbook for the Regional Input-Output Modeling System:* RIMS II. Washington D.C.: U. S. Government Printing Office, May, 1986.

Table 2.1

Period	Average Farm Marketing	Average National Income	National Income Farm Marketing
1921-38	9,478	66,256	6.99
1930-49	14,059	112,568	8.00
1940-59	27,736	264,434	9.53
1950-69	35,468	445,002	12.55
1960-79	63,290	923,671	14.59
1961-80	68,548	1,008,740	14.72

Table 2.2

	National Income		
Period	Originating on Farms	Total	Farm Share
1930-50	8,952	112,568	.0795
1940-60	15,346	264,434	.0580
1950-70	17,043	445,002	.0383
1960-80	26,482	923,671	.0287

FIGURE 1. CONSTRUCTION MATERIALS PURCHASED FOR THE PROPOSED NRP AT SAVANNAH RIVER

INPUT OUTPUT MODEL NUMBER		SECTOR NAME	BLS SEC-TOR	1988/1982 PRODUCER PRICE INDEX	PURCHASES OF MATERIALS ALL ESTIMATES IN 1988 DOLLARS 1989	1990	1991	1992	1993
	69	NEW HIGHWAYS AND STREETS	23	126.1289	12,278	27,706	45,526	81,381	97,244
	72	OTHER CONSTRUCTION/NEW GOV FAC	25	108.5775	402,665	908,668	1,493,066	2,668,989	3,189,246
	74	MAINTENANCE AND REPAIR OTHER FAC	27	121.5192	4,385	9,895	16,259	29,065	34,730
	161	SAWMILLS AND PLANING MILLS, GEN	29	113.4721	9,168	20,690	33,996	60,772	72,618
168M		PRE-FAB WOOD STRUCTURES	34	122.0279	0	0	0	0	0
	181	METAL OFFICE FURNITURE	37	125.4645	4,544	10,255	16,850	30,122	35,993
184M		METAL PARTITIONS, FIXTURES	36	123.9782	0	0	0	0	0
	186	FURNITURE AND FIXTURES, N.E.C.	37	125.4645	11,241	25,368	41,683	74,511	89,035
	192	BUILDING PAPER AND BOARD MILLS	123	116.1306	3,428	7,736	12,712	22,723	27,153
	232	SURFACE ACTIVE AGENTS	137	126.5031	6,059	13,673	22,467	40,162	47,991
234M		PAINTS AND ALLIED PRODUCTS	138	116.4957	0	0	0	0	0
236M		LUBRICATING OILS AND GREASE	142	111.0845	0	0	0	0	0
239M		ASPHALT FELTS, COATINGS	142	111.0845	0	0	0	0	0
258M		BRICK & STRUCTURAL CLAY TILE	41	120.1881	0	0	0	0	0
259M		CERAMIC WALL AND FLOOR TILE	41	120.1881	0	0	0	0	0
	267	CONCRETE BLOCK AND BRICK	40	119.5487	2,950	6,657	10,938	19,553	23,364
	268	CONCRETE PRODUCTS, N.E.C.	40	119.5487	13,952	31,485	51,734	92,478	110,505
	269	READY-MIXED CONCRETE	40	119.5487	1,215,485	2,742,909	4,506,976	8,056,623	9,627,073
	273	ABRASIVE PRODUCTS	41	120.1881	159	360	591	1,057	1,263
	274	ASBESTOS PRODUCTS	41	120.1881	118,313	266,990	438,701	784,217	937,082
280M		BLAST FURNACE, STEEL BOILING MILLS	42	105.9772	0	0	0	0	0
285M		GRAY IRON FOUNDARIES	43	116.279	0	0	0	0	0
289M		DRAWING OF NONFERROUS WIRE	50	96.576	0	0	0	0	0
306M		PLUMBING FIXTURE AND BRASS	55	125.3357	0	0	0	0	0
1/2M	308	FABRICATED STRUCTURAL METAL	56	107.8992	202,743	457,517	751,763	1,343,844	1,605,795
309M		METAL DOORS, SASH, FRAME	56	107.8992	0	0	0	0	0
1/2M	310	FABRICATED PLATE WORK (BOILER SH)	56	107.8992	379,096	855,483	1,405,676	2,512,772	3,002,578
	311	SHEET METAL WORK	56	107.8992	96,389	217,514	357,405	638,894	763,431

FIGURE 2. CONSTRUCTION MATERIALS PURCHASED FOR THE PROPOSED NRP AT SAVANNAH RIVER

INPUT OUTPUT MODEL NUMBER		SECTOR NAME	BLS SEC-TOR	1988/1982 PRODUCER PRICE INDEX	PURCHASES OF MATERIALS FROM AIKEN COUNTY BUSINESS SECTORS ALL ESTIMATES IN 1988 DOLLARS 1989	1990	1991	1992	1993
	69	NEW HIGHWAYS AND STREETS	23	126.1289	9,734	21,967	36,094	64,522	77,099
	72	OTHER CONSTRUCTION/NEW GOV FAC	25	108.5775	370,855	836,884	1,375,116	2,458,142	2,937,299
	74	MAINTENANCE AND REPAIR OTHER FAC	27	121.5192	3,608	8,143	13,380	23,918	28,580
	161	SAWMILLS AND PLANING MILLS, GEN	29	113.4721	8,080	18,233	29,960	53,556	63,996
168M		PRE-FAB WOOD STRUCTURES	34	122.0279	0	0	0	0	0
	181	METAL OFFICE FURNITURE	37	125.4645	3,622	8,174	13,430	24,008	28,688
184M		METAL PARTITIONS,FIXTURES	36	123.9782	0	0	0	0	0
	186	FURNITURE AND FIXTURES, N.E.C.	37	125.4645	8,960	20,219	33,223	59,388	70,965
	192	BUILDING PAPER AND BOARD MILLS	123	116.1306	2,952	6,662	10,946	19,567	23,381
	232	SURFACE ACTIVE AGENTS	137	126.5031	4,790	10,809	17,760	31,748	37,936
234M		PAINTS AND ALLIED PRODUCTS	138	116.4957	0	0	0	0	0
236M		LUBRICATING OILS AND GREASE	142	111.0845	0	0	0	0	0
239M		ASPHALT FELTS,COATINGS	142	111.0845	0	0	0	0	0
258M		BRICK & STRUCTURAL CLAY TILE	41	120.1881	0	0	0	0	0
259M		CERAMIC WALL AND FLOOR TILE	41	120.1881	0	0	0	0	0
	267	CONCRETE BLOCK AND BRICK	40	119.5487	2,467	5,568	9,149	16,355	19,543
	268	CONCRETE PRODUCTS, N.E.C.	40	119.5487	11,671	26,336	43,274	77,356	92,435
	269	READY-MIXED CONCRETE	40	119.5487	1,016,728	2,294,386	3,769,991	6,739,197	8,052,845
	273	ABRASIVE PRODUCTS	41	120.1881	133	299	492	879	1,051
	274	ASBESTOS PRODUCTS	41	120.1881	98,440	222,143	365,012	652,492	779,680
280M		BLAST FURNACE,STEEL BOILING MILLS	42	105.9772	0	0	0	0	0
285M		GRAY IRON FOUNDARIES	43	116.279	0	0	0	0	0
289M		DRAWING OF NONFERROUS WIRE	50	96.576	0	0	0	0	0
306M		PLUMBING FIXTURE AND BRASS	55	125.3357	0	0	0	0	0
1/2M	308	FABRICATED STRUCTURAL METAL	56	107.8992	187,900	424,023	696,727	1,245,463	1,488,236
309M		METAL DOORS, SASH, FRAME	56	107.8992	0	0	0	0	0
1/2M	310	FABRICATED PLATE WORK (BOILER SH)	56	107.8992	351,343	792,854	1,302,768	2,328,814	2,782,762

FIGURE 3A. U.S. MANUFACTURED TOBACCO EXPORTS AND ASSOCIATED TRADE MARGIN VALUES, MILLION DOLLARS, 1987

		INDUSTRY SALES BY I/O SECTOR NUMBER						
PRODUCTS	FAS* EXPORT VALUE	RAW TOBACCO 15	CIGARETTES 127	CIGARS 128	CHEWING & SMOKING TOBACCO 129	TOBACCO STEMMING & REDRYING 130	TRANSPOR-TATION 446	WHOLESALE TRADE 460
	1982 PRICES, MILLION DOLLARS							
RAW EXPORTS	0.00	0.00	0.00	0.00	0.00	0.00	0.00	0.00
CIGARETTE EXPORTS	1534.76	0.00	1473.37	0.00	0.00	0.00	15.3476	46.0428
CIGAR	6.58	0.00	0.00	6.3168	0.00	0.00	0.0658	0.1974
CHEWING	185.68	0.00	0.00	0.00	178.2528	0.00	1.8568	5.5704
STEMMING	985.55	0.00	0.00	0.00	0.00	798.2955	68.9885	118.2660
TOTAL		0.00	1473.37	6.3168	178.2528	798.2955	86.2587	170.0766

*FAS – free along-side ship.

FIGURE 3B. THE ECONOMIC IMPACT OF U.S. MANUFACTURED TOBACCO EXPORTS, MILLION DOLLARS, 1987

	SECTOR	FINAL DEMAND	TGO	EMPLOYEE COMP INCOME	PROPERTY INC.	TOTAL INCOME	VALUE ADDED	EMPLOY-MENT
		----------1982 PRICES, MILLION DOLLARS----------						(NO. JOBS)
1	AGG AGRICULTURE	23.0448	189.8567	13.9786	43.9555	57.9341	61.4100	2776.96
15	TOBACCO	0.0052	796.8186	81.8754	346.0865	427.9619	444.7999	17448.06
26	AGG AG SERVICES	0.5808	29.8229	11.5509	4.8390	16.3899	17.0301	1249.76
28	AGG MINING	4.4634	253.1472	31.7751	94.1575	125.9326	150.8737	846.78
66	AGG NEW CONST	0.0000	0.0000	0.0000	0.0000	0.0000	0.0000	0.00
73	AGG MAINT CONST	0.0000	140.4349	59.2348	5.9913	65.2261	67.0116	2229.03
76	AGG ORDNANCE	0.9523	1.7239	0.6916	-0.4623	0.2293	0.2484	19.10
82	AGG FOOD PROCESS	188.6762	322.9005	43.0684	26.8393	69.9077	78.6938	1831.71
127	CIGARETTES	1492.3580	1518.4510	148.6240	317.8269	466.4509	704.4136	4356.29
128	CIGARS	6.6646	6.8145	3.0083	0.3462	3.3546	4.0966	143.72
129	CHEWING AND SMOKING TOBA	178.9208	183.6485	28.8225	32.0259	60.8484	63.6081	1427.71
130	TOBACCO STEMMING AND RED	798.2973	1263.6380	49.1260	57.6129	106.7389	108.7991	1661.17
131	BROADWOVEN FABRIC MILLS	1.9413	31.7485	6.5364	1.1821	7.7185	7.9789	376.68
132	AGG TEXT,WOOD PROD	72.8038	144.0480	37.4603	10.3764	47.8367	49.0101	2432.28
188	PAPER MILLS, EXCEPT BUIL	0.3076	70.2923	16.6992	4.3636	21.0628	22.4041	492.59
189	AGG OTHER PAPER	10.1342	51.2703	10.8543	4.2482	15.1025	15.7467	400.67
193	PAPER COATING AND GLAZIN	0.3154	20.5386	4.6093	2.1314	6.7407	6.9955	198.69
199	PAPERBOARD CONTAINERS	0.2966	45.7018	9.7760	2.9627	12.7388	13.3092	382.34
200	AGG OTHER PRINTING	19.7565	141.2753	43.8092	13.5611	57.3703	58.6839	1923.47

Note: AGG denotes aggregate.

FIGURE 4. TOTAL MULTIPLIERS, BY INDUSTRY AGGREGATION, FOR OUTPUT, EARNINGS, AND EMPLOYMENT

	OUTPUT[1] (DOLLARS)	EARNINGS[2] (DOLLARS)	EMPLOYMENT[3] (NO. OF JOBS)
AGRICULTURE, FORESTRY, AND FISHERIES:			
AGRICULTURAL PRODUCTS AND AGRICULTURAL, FORESTRY, AND FISHERY SERGVICES	1.8934	0.5412	74.3
FORESTRY AND FISHERY PRODUCTS	1.6651	0.3121	36.2
MINING:			
COAL MINING	1.6558	0.5014	47.3
CRUDE PETROLEUM AND NATURAL GAS	1.4646	0.2273	15.1
MISCELLANEOUS MINING	1.7991	0.5165	31.4
MANUFACTURING:			
FOOD AND KINDRED PRODUCTS AND TOBACCO	2.1428	0.4401	39.9
TEXTILE MILL PRODUCTS	2.7121	0.6382	43.5
APPAREL	2.6766	0.7902	57.0
PAPER AND ALLIED PRODUCTS	2.1521	0.5223	27.9
PRINTING AND PUBLISHING	1.9635	0.6180	41.4
CHEMICALS AND PETROLEUM REFINING	2.0123	0.4696	23.9
RUBBER AND LEATHER PRODUCTS	2.0094	0.5225	28.5
LUMBER AND WOOD PRODUCTS AND FURNITURE	2.3169	0.5778	43.0
STONE, CLAY, AND GLASS PRODUCTS	2.0217	0.6021	34.1
PRIMARY METAL INDUSTRIES	1.7852	0.4031	20.8
FABRICATED METAL PRODUCTS	1.8268	0.5004	28.9
MACHINERY, EXCEPT ELECTRICAL	1.8913	0.5649	31.8

1. Each entry in column 1 represents the total dollar change in output that occurs in all row industries for each additional dollar of output delivered to final demand by the industry corresponding to the entry.

2. Each entry in column 2 represents the total dollar change in earnings of households employed by all row industries for each additional dollar of output delivered to final demand by the industry corresponding to the entry.

3. Each entry in column 3 represents the total change in number of jobs in all row industries for each additional 1 million dollars of output delivered to final demand by the industry corresponding to the entry.

Source: Regional Input-Output Modeling System (RIMS II), Regional Economic Analysis Division, Bureau of Economic Analysis.

3

The Problem: Using Value-Added Information in Benefit/Cost Analysis

Stephen C. Cooke

Introduction

The change in value added would seem to be an alluring way to gauge the general equilibrium benefits from a public investment project. Young and Gray list the reports using value added to measure net benefits (p. 1819). The use and abuse of value added in benefit/cost analysis has been the focus of attention in several recent articles in the regional economics literature (Hamilton and Gardner, Stabler, et. al. and Young and Gray). A key question posed by the authors in each instance is whether value added is ever an acceptable measure of net benefits in a benefit/cost analysis of a publicly supported irrigation project? Another is whether the benefits in a benefit/cost analysis are measured differently when the economy under consideration is at full-employment or less-than-full-employment? The controversy focuses on benefits. Typically, the costs of a public project can be determined relatively easily and accurately, since they tend to appear ex post as an item in some government agency's budget.

In this paper, it is argued that economic surplus is the only appropriate measure of benefits for public investment.[1] However, there is still a very important role that value-added information can play in the process of determining economic surplus. Ex post, value-added information can be used to determine the change in productivity that resulted from a public investment. A productivity index is a necessary but not sufficient information for measuring economic surplus.[2] Therefore, value-added information is indirectly helpful in measuring the benefits to public investment in a full-employment or "taut" economy. It is hoped that some light can be shed on the role of value added in measuring economic surplus in the less-than-full-employment or "slack" economy as well.

The Relationship Between Factor Intensity and Productivity

The benefits from a public investment in a sector represent a positive-sum increase in the real income, as measured by the change in economic surplus, for one or more groups of people. Given constant demand, economic surplus increases when the production of a commodity can be increased with a less than proportional increase in the quantity of inputs, i.e., increased productivity. In such cases, public investment in human and physical capital causes a rise in productivity of the sector that increases value added and economic surplus. Value added is the payment to the primary factors of capital and labor. An increase in value added, as demonstrated below, is equal to the increases in capital and labor intensity and productivity. Changes in economic surplus, on the other hand, are equal to the increases in real income to consumers and resource owners.[3] The greater the increase in productivity, the larger the economic surplus and the more efficient the public investment. Solow has shown that continuing growth in productivity (and continuing public investment) can result in stable growth in real income over time (Solow, p. 38).[4]

Increasing labor and/or capital intensity -- as opposed to productivity -- also increases output and value added, at least temporarily. It can be shown that additional capital investment in a full-employment economy, without an accompanying increase in productivity, results in a temporary increase in real income only and is*not* a defensible justification for public investment. Increasing the rate of growth in capital investment beyond the "natural rate" i.e., the rates of growth in the labor supply and in technological change, will not result in a further increase in the rate of growth in real income in the long run (Solow, p. 38). Up to the natural rate of investment, capital is needed to complement increases in the labor supply and in productivity.

Solow illustrates this phenomenon with the following "stylized facts" (Solow pp. 23-30). Assume there is an increase in savings and investment (or taxes and government spending) without any associated increase in productivity. Implicitly, this assumes that the additions to the stocks of physical and/or human capital are homogeneous -- just "more of the same." Assume also that, at a constant rate of unemployment, the supply of labor is fixed. Then additional capital investments increase the ratio of capital to labor, i.e., increase capital intensity. Output per worker increases and output per unit of capital decreases. The payoff is a permanently higher savings rate and a lower consumption rate per capita (Solow, pp. 23-24).[5] The high Japanese saving rate might make that economy an example of this phenomenon.

When technological change is included in the model, the equilibrium growth rate of savings and investment in capital expands beyond the rate of growth in the labor force and now includes the rate of growth in productivity as well.

> The natural rate of growth [in capital and output] is . . . the sum of the rate of population increase and the rate of technological progress. A change in the savings rate does not change that; . . . an increase in the rate of technological progress itself, besides increasing the rates of growth of output and output per head (therefore consumption per head), will also increase effective employment per unit of capital (Solow, p. 38).

Increased Productivity and Changing Economic Surplus

When public investment increases productivity, it increases the marginal productivity of each input, which shifts the supply curve outward thereby increasing economic surplus. Thus, the public investment project that increases economic surplus affects different categories of people either positively and negatively. Consumers benefit from a decrease in the commodity price, which is measured as consumer surplus. Resource owners of inputs available in less than perfectly elastic supply benefit from an increased input price, which is measured as economic rent. Economic surplus is the sum of consumer surplus and economic rent.

The groups negatively affected by an increase in productivity include those who own (or, in the case of labor, are) "saved resources" that have low opportunity costs and become underemployed elsewhere in the economy. These people experience the full fury of Schumpeter's "gales of creative destruction." The measure of economic surplus assumes saved resources are re-employed at their ex ante opportunity cost. Saved resources that are re-employed below their previous factor price represent a restructuring cost that must be addressed when estimating the change in economic surplus. In fact, there are two restructuring-of-underemployed-resources issues that need to be consistently resolved. First, there is the problem of whether to count as an additional cost, saved resources that become underemployed as a result of the public investment. Second, there is the issue of whether previously underemployed resources that are re-employed because of a given public investment should be counted as an additional benefit. Both of these problems will be discussed subsequently. For now, the assumption of a taut economy precludes both of these questions.

We need a measure of the shift in the supply curve associated with the change in productivity and a measure of the change in economic surplus associated with the change in real income. The net change in economic surplus is the measure of benefits then used in a benefit/cost calculation. To measure economic surplus ex post, as a parallel shift in the supply function, the equations suggested by Rose can be used (Rose, p. 834-5).

$$(1)\quad ES = \tfrac{1}{2}kP_0(Q_0 + Q_1)$$

where ES is economic surplus; k is an index of the shift in the supply curve, i.e., total factor productivity; P_0 is the initial commodity price; Q_0 is the initial commodity quantity; and Q_1 is the subsequent quantity.

Equation (1) is used to estimate economic surplus. This equation requires measures of P_0, Q_0, and Q_1 that are either known or readily determined ex post. Equation (1) also requires, not surprisingly, a measure of the shift in supply k that is associated with the change in productivity and remains to be derived.

To summarize, only economic surplus or the sum of consumer surplus and economic rent are used to measure benefits in a full-employment benefit/cost analysis. The measure of economic surplus requires a measure of the shift in supply associated with a change in productivity. Value added can be use to measure the change in total factor productivity that causes the supply shift. Value added is *not* a direct measure of economic surplus but it can be used as a direct measure of productivity. In the next section, the change in value added and its components will be used to measure productivity as a measure of the k shift in the supply curve needed in equation (1).

Total Factor Productivity and Value Added

Consider a continuous, twice differentiable, concave, linearly homogeneous production function for industries in a full-employment general equilibrium economy in which output is a function of primary inputs and a discrete variable for time.

$$(2)\quad V_i = F(K_i, L_i, T_i)$$

where V is the output in terms of value added in sector i; K and L are capital and labor in sector i; and T is time used as a discrete measure of technological change.[6]

Equation (2), expressed in terms of growth, is

$$(3)\quad G_{Vi} = G_{Ti} + S_{Ki}G_{Ki} + S_{Li}G_{Li}$$

Where G_{Vi} is the rate of growth in value added in sector i, G_{Ki} is the rate of growth in the quantity of capital, G_{Li} is the rate of growth in the quantity of labor, and G_{Ti} is the rate of growth in total factor productivity, S_{ji} is the share of value added of primary input j (Chenery, p. 17).[7]

Solving for the growth in productivity,

$$(4) \quad G_{Ti} = G_{Vi} - S_{Ki}G_{Ki} - S_{Li}G_{Li}$$

Equation (4) indicates that the growth in total factor productivity equals the growth in value added less the growth in capital and labor weighted by their factor shares. Thus, the growth in value added over-estimates the growth in total factor productivity by the growth in capital and labor intensity. This result is consistent with similar conclusions reached by Young and Gray, Hamilton and Gardner, and Stabler et al. "It is obvious . . . that input-output models are poorly suited for calculating indirect benefits, properly defined. Optimally what is required is a computable general equilibrium model" (Stabler et al. p. 13). It is argued here that value-added information (including that found in two independently derived input-output models of a region) can be used to measure the general-equilibrium supply shift as an increase in productivity that is needed to estimate benefits, both direct and indirect.

Equation (4) can be expressed more generally for empirical purposes in terms of a non-homothetic production function expressed in logarithms. Diewert's quadratic lemma is then applied to determine a second-order approximation of the change in productivity as the geometric mean of V_a and V_b expanded around points in time a and b (Diewert, p. 118). Dropping the i subscript,

$$(5) \quad ½(\alpha_a + \alpha_b)\, T_a\text{-}T_b = \ln V_a\text{-}\ln V_b - ½(S_{Ka}+S_{Kb})\,(\ln K_a\text{-}\ln K_b) - ½(S_{La}+S_{Lb})\,(\ln L_a\text{-}\ln L_b),\ (5)$$

where α_t is a measure of productivity relative to time t. The information needed to estimate equation (5) includes initial and subsequent value added (V_a, V_b), quantity of capital (K_a, K_b) and labor (L_a, L_b), and factor shares of capital (S_{Ka}, S_{Kb}) and labor (S_{La}, S_{Lb}). Equation (5) measures the index of total factor productivity as the difference between the index of output and a Tornquist index of inputs, adjusted for changing factor prices (Diewert, p. 120).[8] The left side of equation (5) is a measure of the k shift in the supply curve referred to in equation (1) that is used to determine economic surplus.

Measuring Benefits and Costs in a Slack Economy

Is there any difference in measuring the benefits and costs to public investments in a slack economy compared to a taut one? If there are underemployed resources in a region, then it is reasonable to expect that the change in economic rent will be less since it will be easier to entice resources from alternative employment with a minimal increase in factor price, i.e., supply will be more elastic. Also the project costs should be reduced since the price of the needed capital and labor inputs will have lower opportunity cost.[9]

When measuring benefits in a slack economy, we must also address the issues of dealing consistently with the problems of (1) saved resources that become underemployed and (2) underemployed resources that are re-employed for a given public investment. On the general issue of including changes in slack as a measure of economic growth Solow states:

> One of the contributions of the modern theory of growth has been to put a damper on loose discussion of policy directed to change the rate of growth. The year-to-year growth of real output in an economy has three elements. Some of it comes from year-to-year changes in the degree of utilization or slack in the economy, as measured by unemployment rate or rate of capacity utilization. An economy can grow faster or slower from one year to the next because its unemployment rate is falling or rising. If this is to be described as growth, it is specifically growth of demand, not growth of supply. Growth in supply, or productive capacity, has two further components. One is the underlying steady-state rate of growth, the natural rate, the other is the growth that comes from a current or recent change in the proportion of output invested. The theory says that this last component of growth is transitory (Solow, p. 78).

Solow suggests that it is the responsibility of the government, through its fiscal and monetary policy, to maintain a "fairly steady" rate of employment. Solow also recommends that the criterion for public investment should be "to keep the marginal product of industrial capital equal to the marginal product of overhead capital at every instant of time" (Solow, p. 94). This suggests that the federal government's responsibility to maintain full-employment by shifting aggregate demand is separate from its (or other governmental units') responsibility to make investments in infrastructure based on marginal productivity and measured as benefit/cost or internal rate of return. This implies that the re-employment of underemployed resources from infrastructure investment should not count as an additional benefit of a government's supply-side investment policies beyond those outlined above.[10] Similarly, saved resources that become underemployed as a result of infrastructure investment should not be subtracted as a cost.[11] In both cases, the change in underemployment is properly a measure of the performance of the federal government's fiscal-monetary policies. To include the effects of underemployed or immobile resources of a slack economy into benefit/cost analysis is tantamount to weighting a long-run investment measure by the effect of the federal government's short-run fiscal and monetary policy on structural adjustments. Therefore, it is concluded that the procedure for measuring the direct and indirect benefits from investments in infrastructure as the change in economic surplus should be the same in the context of either a taut or slack economy.

Summary and Conclusions

The benefits in a benefit/cost analysis are measured by changes in economic surplus. An increase in total factor productivity will increase economic surplus. Increases in homogeneous capital do not increase productivity. However, value-added information, including the change in capital and labor intensity, can be used to determine the change in productivity that is a necessary component of economic surplus.

It is shown that the change in value added less the changes in capital and labor intensity equals total factor productivity. It is important to note that total value added over-estimates changes in productivity since it also includes changes in capital and labor intensity. Thus, by itself, value added is not the measure of benefits needed for a benefit/cost analysis. However, the growth in value added is a necessary component in determining the change in total factor productivity. Productivity is necessary information in estimating economic surplus as the measure of the benefits accruing from public investments. Diewert's quadratic lemma can be used to compute an ideal index of productivity from information on value added.

It is also concluded that economic surplus is the appropriate measure of benefits in either a taut and slack economy. Nonetheless, the temptation to use value added directly as a measure of benefits can be great. Especially for a poor community, the secondary effects from using unemployed and under-employed labor and capital seductively suggests using adjusted value added as a measure of benefits to public investment. If employment increases as capital investment increases, then real growth in value added will occur until a fixed rate of unemployment is reached. To some, this may seem to justify using an adjusted value added to measure the benefits to public investment in a depressed region, although a true measure of the benefits is an estimation of the change in economic surplus. These instances serve to point out the pressure an analyst can come under to incorrectly use value added as a direct measure of benefits from a public investment.

Notes

[1] In the literature, this point of view has also been adopted by Young and Gray and by Stabler et al: ". . . the economic surplus approach developed in applied welfare and economics should be the guiding criterion in regional, as in

national contexts" for benefit/cost analysis (Young and Gray, P. 1820); and ". . . value added is not a measure of benefit; it is merely the upper limit on the opportunity cost of the resources employed . . ." (Stabler, et al, p. 16).

[2] Again, this point has been recognized, at least implicitly, by Young and Gray and by Stabler et al: ". . . for a water resource project to leave the region better off than it was without the investment, the value of the incremental output must exceed the sum of the opportunity costs of all resources required for the public development of the project and for the private utilization of the water" (Young and Gray, p. 1822). Euler's theorem would suggest that this is only possible for a linearly homogenous production function when increases in productivity take place. Also, "the irrigation project is modelled as an outward rotation of the supply curve for agricultural products" (Stabler, et al, p. 17).

[3] An increased payment to a primary factor may or may not include economic rent depending on the elasticity of supply of the input faced by the firm, industry, or sector. The total payment to a primary factor may increase simply because more units are needed at a given price. Economic rent refers to the situation in which the factor price on all units must be bid up in order to entice the marginal units from their employment elsewhere in the economy.

[4] Solow's neoclassical model of growth assumes that the economy is in competitive equilibrium such that there is a Pareto optimal allocation of resources, with each sector making equally productive use of the capital and labor available. It is likely, however, that there are systematic variations in the returns to labor and capital in different sectors. These variations would make it possible to increase output by reallocating capital and labor from less productive to more productive sectors. Therefore, other sources of growth in output include reallocating resources between sectors, economies of scale, and reduction of internal and external bottlenecks (Chenery, p. 15). In this paper, productivity is defined to include changes in total factor productivity and scale economies only. "Productivity gains" from resource allocation and reduction of bottlenecks are associated with the restructuring of a "slack" economy.

[5] "Initially, the rate of growth of output must be higher than the steady-state rate of growth. . . . But eventually the economy approaches its new steady state; the rate of growth of output slows down to the rate of growth in the labor force . . . (Solow, p. 26). Consumption is maximized when the marginal product of capital is just equal to the rate of growth in the labor force" (Solow, pp. 27-28).

[6] Value added is a function of capital and labor only, since the inclusion of intermediate inputs would amount to double counting in a general equilibrium context (Chenery, p. 17). For partial equilibrium analysis, gross industrial output is a function of both primary and intermediate inputs.

[7] This equation is derived from a Cobb-Douglas production function.

[8] This measure of total factor productivity will include the effects of changes in economies of scale since no steps have been taken to separate these two sources of productivity gain.

[9] Mishan states that ". . . the advantages of public investment in times of low employment are made manifest by reference to the cost aspect. For where there is substantial unemployment, the opportunity cost of labor, skilled and unskilled, and indeed of specific forms of capital equipment, is much lower than if such factors are already employed. Thus investment projects that would not be economically feasible under conditions of full employment -- assuming, of course, that employment is expected to remain low, at least in the absence of these investments" (Mishan, p.293).

[10] Mishan takes the opposite point of view regarding secondary effects. However, he implies that the strongest case for including secondary effects can be made only under a narrowly defined set of circumstances. ". . . cost benefit calculations that take no account of . . . secondary income and employment effects will underestimate the net benefits of the projects involved. Although allowance for these secondary income effects should obviously be made -- at least wherever, under existing political circumstances, no alternative ways of expanding employment are anticipated -- we shall restrict ourselves . . . to the primary employment effects" (Mishan, p. 294).

[11] This result contradicts that reached by Schmitz and Seckler. "In order to determine the value of the harvester, we have to determine whether the gainers (producers, consumers, etc.) could compensate the losers (workers) and still be better off than before (Schmitz and Seckler, p. 574). And ". . . compensation is a necessary but not a sufficient condition for appraising an improvement (Schmitz and Seckler, p. 575, fn 11). Since economic surplus is equal to resources saved, the Schmitz and Seckler criterion reduces to a question of opportunity cost and asset mobility. Beyond this, it is assumed that separate economic measurements of efficiency and distribution can be made and therefore should not be lumped together. The choice between equity and efficiency is ultimately a political one.

References

Chenery, H. (1986) "Growth and Transformation." *Industrialization and Growth: A Comparative Study*. Ed. by H. Chenery, S. Robinson, and M. Syrquin. Oxford University Press, New York.

Diewert, W. E. (1976) "Exact and Superlative Index Numbers." *Journal of Econometrics* 4:115-145.

Hamilton, J. and R. Gardner. (1986) "Value Added and Secondary Benefits in Regional Projection Evaluation: Irrigation Development in the Snake River Basin." *The Annals of Regional Science* 20(1): 1-11.

Mishan, E. J. (1976) *Cost-Benefit Analysis*. Praeger Publishers, New York.

Rose, R. (1980) "Supply Shifts and Research Benefits: Comment." *The American Journal of Agricultural Economics* 62(4):834-837.

Schmitz, A. and D. Seckler. (1970) "Mechanized Agriculture and Social Welfare: The Case of the Tomato Harvester." *American Journal of Agricultural Economics* 52:569-77.

Solow, R. (1970) *Growth Theory: An Exposition*. Oxford University Press, New York.

Stabler, J., G. Van Kooten and N. Meyer. (1988) "Methodological Issues in the Evaluation of Regional Resource Development Projects." *The Annals of Regional Science* 22(2):13-25.

Young, R. and S. Gray. (1985) "Input-Output Models, Economic Surplus, and the Evaluation of State or Regional Water Plans." *Water Resources Research* 21(12):1819-1823.

4

A Survey Approach to Developing an Input/Output Model

Michael W. Babcock

Introduction

Input-output analysis is one of the most durable methods in the field of regional science. Despite well-known theoretical and empirical problems, input-output continues to thrive and grow. Perhaps this is due to the flexibility and descriptive power of input-output analysis. The transactions table tells us how the regional economy works. It describes how the regional industries interact with each other and with the outside world, through imports and exports. Input-output models are capable of simulating almost any conceivable economic impact. Once the analyst has measured final demand changes due to the economic impact, the model will measure the impact on total output and the output of every regional industry. Of course this advantage also applies to regional economic forecasting.

Input-output also has disadvantages. The assumptions of perfectly elastic supply and fixed technical coefficients constrain the practical uses of the method. That is, input-output is most likely to generate believable results when used to measure modest economic change of a short run nature. However the main problem of input-output is the high time and money costs of obtaining the data.

All regional input-output models can be placed into one of three classes. These are survey, non-survey, and hybrid models. The models differ in the extent to which they use primary or secondary data sources. Survey based models obtain most of the data for the transactions table through mailed questionnaires or personal interviews of regional business firms. Reliance on secondary data is usually limited to developing control totals, filling in blank cells, and reconciling differences between purchase and sales data in specific cells of the table. Non-survey models employ almost no primary data and usually obtain regional data by adjusting national input-output data.[1] Hybrid models usually rely on surveys to obtain the largest regional I-O coefficients and secondary data for the rest of the table.[2]

The objective of this paper is to provide a blueprint to implement a survey based regional I-O model.

Blueprint for a Survey-Based Regional I-O Model

If done well, survey-based I-O models have potential to be much more accurate than other approaches. Relatively few survey based studies exist (Table 4.1.), and as far as I know, only the 1985 Kansas study has been conducted since 1975. This is due to alleged high time and money costs of the survey approach. However the money costs are not that high. The budget for the Kansas study was about $125,000 which is rather small for sponsored research projects.

The remainder of the paper is a suggested blueprint for survey based regional I-O models, based on my experience as co-investigator in the 1985 Kansas study.

Study Area

The first step is to define the study area. This is usually a non-issue. Despite the theoretical elegance of the economic region concept, the study area is usually defined by the granting agency. This is reinforced by the fact that secondary data is classified by governmental jurisdiction. Survey based models have been conducted for states, MSA's, and counties.

Objectives

The second step is definition of research objectives. This depends a great deal on the motivation of the granting agency. A clear statement of objectives is necessary before doing anything else since objectives determine the transactions table. If the objective is to stimulate regional exports and minimize imports, the table will require some disaggregation of interregional flows. On the other hand, if the objective is to attract new business, the emphasis would be on intraregional linkages.

Secondary Data

The third step is a secondary data search. The classification of regional I-O models into survey and non-survey groups has always been artificial since good secondary data is essential to a successful survey study. Secondary data serves a variety of useful purposes. These include industry control totals, filling in cells where survey data is unavailable or unreliable, and reconciling conflicting purchase and sales data in specific cells. For some of the final demand sectors, such as the government sectors, there is really no alternative to secondary data. Perhaps the most significant benefit of secondary data is the description it provides of the region's industrial structure. By revealing the region's industrial specialization, the researcher discovers which industries will require intensive research effort and which ones do not. The secondary data search will also give the researcher a good idea of how the final transactions table should appear.

Population

The fourth step is to obtain the population of firms in the study area. Firms cannot be surveyed if researchers are unaware of their existence. The researcher also needs the population in order to draw representative samples. Every business firm has to file a tax return. Therefore state agencies that collect taxes have lists of firms doing business in the state. The Kansas study was greatly assisted by an excellent Directory of Manufacturers published by the Kansas Department of Commerce. It provided everything we needed to know since all Kansas manufacturing firms were classified by SIC, primary products, employment size, and location. It also contained phone numbers and addresses of every manufacturing firm.

Sectoring

The fifth step is to decide which sectors will be used to specify the transactions table. The general principle with respect to sectoring is homogeneity. Firms with similar product mixes and production functions should be included in the same sector. Sometimes the principle of homogeneity is qualified by practical considerations. For example, industries that play a minor role in the regional economy can be aggregated into a few non-homogeneous sectors to minimize data collection costs. This is also true of innocuous expenditures like "office supplies."

Other factors also influence sectoring. The size of the project budget has a major impact. The bigger the budget the greater the degree of disaggregation. The need to avoid disclosure of individual firm data is also significant. Perhaps the most important factor is the need to accurately represent the industrial structure of the region. The more important industries should have the greatest level of disaggregation. For example, Kansas is known for its food and agriculture as well as its transportation equipment industry. Of the 69 sectors in the 1985 Kansas I-O study, 20 pertain to food and agriculture and 8 to the transportation equipment industry.

Control Totals

The sixth step is development of control totals of gross output for each sector in the transactions table. Control totals obtained from secondary data play a variety of roles in a survey based study. In the few sectors where 100 percent coverage is achieved, the control totals are used to check reliability. However most sectors contain data obtained from a sample of the region's firms. In these cases, control totals are used to "blow up" the sample into transactions for the entire sector.

Sometimes it is necessary to use regional control totals with national coefficients to estimate cells where survey data is unreliable or unavailable. The various federal government censuses are excellent sources of regional control data. A list of these is in Appendix A.

Select Industry Samples

The seventh step is selection of survey samples for each sector. The general principle is to concentrate research effort on the large firms in the important industries, simply because these are the firms that account for the largest flows. In these cases the researcher should strive for 100 percent coverage. For example the Kansas airplane manufacturing industry consists of Boeing, Cessna, Beech and Gates Learjet. If the researcher leaves any of these firms out of the sample it puts a big hole in the analysis. The Kansas automobile manufacturing industry is one firm--the GM plant in Kansas City, Kansas. If GM's data is not obtained, there will not be an automobile manufacturing sector. The same principle is true of Kansas meatpacking where IBP and Excel dominate the industry.

In the less concentrated sectors, the researcher should draw a stratified sample including both large and small firms so that both are represented. Firms that refuse to cooperate should be replaced by another firm of similar size if possible.

The Questionnaire

A questionnaire is necessary whether the data is obtained by mail or personal interview. The general principles are brevity and clarity. It should have lots of explanatory notes and definitions of terms. The questionnaire for the 1985 Kansas study has three pages (see attached questionnaire on 67 - 69). The first page pertains to input purchases and other costs. The respondent lists the name of each supplying sector, the total purchases from that sector, and the percent supplied by firms in Kansas and the respondent's county. The difference between total purchases and purchases from Kansas suppliers is imports. The purchase page also contains the typical costs of every firm including wages and salaries, taxes and depreciation.

The second page pertains to the firm's interindustry markets and final demand. The respondent lists the market, the total sales to that market, and the percent of sales made in Kansas and the respondent's county. The residual in each case is exports. The sales page also contains the other final demand sectors which include consumption, state government, local government, federal government (non-defense), and federal government (defense).

The third page contains the sector names which we hoped the respondents would use in describing their purchases and sales. If none of these sector names applied, the respondent was instructed to describe the purchaser or sale as clearly as possible.

The Survey

Data for the 1985 Kansas I-O study was collected by mailed questionnaire (small firms) and personal interviews (large firms). About eight graduate and undergraduate students participated as interviewers. Training sessions were held prior to any interviewing. The sessions included instruction in I-O analysis, how to respond to likely questions, and how to act and dress.

The first step in the survey process is to send a letter to the firm's controller explaining the project. The letter should contain the public policy objectives of the study and how the project could benefit the firm. For example point out that the information being collected will assist economic development which will mean lower cost for local suppliers and more local markets.

The next step is to call the controller (they have the data) and discuss the project. Explain the project fully, *emphasize confidentiality*, and ask for an appointment.

At the interview explain the questionnaire in detail and fully answer all questions. Ask the respondent to describe each supplier and market clearly. Tell them not to worry if the name is not on the sector list (these can be placed in the correct sector after you get the data). Take your time, talk about what the respondent is interested in, tour the plant if offered. All these things inspire confidence and trust. Remember they are doing you a favor.

Leave the questionnaire with the controller and ask to get it back in a couple of weeks. If they cannot complete within two weeks, give them a month. The entire process is a negotiation. Be persistent. Keep calling and eventually most of the respondents will give you the data.

After receiving a completed questionnaire, check it for obvious errors, inconsistencies, and omitted data. Call the controller to clarify these.

Preparation of the Transactions Table

The data for the transactions table in the Kansas study was tabulated on a personal computer. As completed questionnaires were returned, the data was added to the computer file. The personal computer was also used to calculate all inverses. Calculating the inverse for the 69 sector Kansas model took 2.5 minutes.

Several conventions were followed in preparing the transactions table including:

1. Firms were assigned to sectors according to their principal product even though this sometimes violated the homogeneity principle.
2. Sales of the capital goods industries were shown as sales to Gross Private Investment rather than the industry making the purchase.
3. Interindustry flows were measured in producers' prices (i.e. f.o.b. prices). This avoids double counting of transportation costs.
4. The output of wholesale and retail trade as well as most services is defined as gross margin or roughly gross sales minus cost of goods sold.

Consistency Check

The final step in the process is the consistency check. In some cases, specific cell entries may diverge. That is, the row (sale) entry may be different from the column (purchase) entry. The researcher has to reconcile these differences using all available data and best judgment as to which number is more reliable.

The I-O table is a logically consistent accounting framework, making it possible to detect errors and inconsistencies. If large adjustments to individual cells are necessary to achieve accounting balance you know you have major errors.

Conclusion

After two years of hard work by a dozen people and expenditure of $125,000, the question must be asked, "was it worth it?"[3] Of course different people would have different answers to this question. Do the costs outweigh the benefits of having data you know are accurate? Richardson has argued that survey based models are too expensive, non-survey methods are inaccurate; and therefore the future of regional I-O research is the development of hybrid or "mongrel" models that combine some primary data with the best available secondary data.[4] He may be right. However there is still a role for "pure survey" studies if for no other reason than to give non-survey researchers a method of evaluating their research. This paper provides a framework, based on our experiences in Kansas, to guide researchers in constructing a survey based I-O model.

Notes

1 For a comprehensive discussion of non-survey methods see Richardson 1985, 618-630.

2 Ibid. pp. 624-626.

3 After completion of the first Kansas survey based I-O model in 1969 Jarvin Emerson was quoted as follows: "One can simultaneously complete two types of survey based regional I-O models -- your first and your last!"

The costs of the Kansas I-O model are much less than those of a research team with no regional I-O experience. Dr. Emerson has devoted most of his professional career to the study of primary data regional I-O models. Dr. Babcock also has considerable experience in gathering primary data in connection with rural transportation studies.

4 Richardson 1985, p. 609.

References

Alward, G. S. and C. J. Palmer. *IMPLAN: An Input-Output Analysis System for Forest Service Planning.* Fort Collins: U.S. Forest Service, 1981.

Bourque, P. J. and R. S. Conway, Jr. *The Washington Input-Output Study.* Seattle: University of Washington, Graduate School of Business Administration, 1977.

Bradley, I. E., J. H. Short, and F. R. Kolb. *Utah Input-Output Study: Projections of Income, Employment, Output, and Revenue.* Salt Lake City: University of Utah, Bureau of Business and Economic Research, 1970.

Brucker, S. M., S. E. Hastings and W. R. Latham. "Regional Input-Output Analysis: Comparison of Five 'Ready Made' Model Systems," *The Review of Regional Studies,* 17 (1987), 1-15.

Bureau of Business Research, University of New Mexico. *A Technical Supplement to the Input-Output Study for New Mexico.* Albuquerque: University of New Mexico, 1967.

Emerson, M. J. *The Interindustry Structure of the Kansas Economy.* Topeka: State of Kansas, Office of Economic Analysis, 1969, updated 1985.

Emerson, M. J. and J. D. Reed. *The Interindustry Structure of the Wichita Economy.* Topeka: Kansas Office of Economic Analysis, 1971.

Gamble, H. B. and D. L. Raphael. *A Microeconomic Analysis of Clinton County, Pennsylvania.* University Park: Pennsylvania State University, 1965.

Gamble, H. B. *The Economic Structure of Sullivan County, Pennsylvania.* University Park: Pennsylvania State University, Bulletin 743, 1967.

Hirsch, W. Z. "Interindustry Relations of a Metropolitan Area," *Review of Economics and Statistics*, 4 (1959), 360-369.

Hochwald, W., H. E. Striner, and S. Sonenblum. *Local Impact of Foreign Trade.* New York: National Planning Association, 1960.

Isard, W., T. W. Langford, Jr. and E. Romanoff. *Philadelphia Region Input-Output Study.* Philadelphia: Regional Science Research Institute, Working Papers, 1966-68.

Kokat, R. G. "Some Conceptual Problems in the Implementation of the Maryland Interindustry Study," *Papers, Regional Science Association*, 17 (1966), 95-104.

Miernyk, et al. *Simulating Regional Economic Development: An Interindustry Analysis of the West Virginia Economy.* Lexington, Mass: Lexington Books, 1970.

Nevin, E. T., A. R. Roe, and J. I. Round. *The Structure of the Welsh Economy.* Cardiff: University of Wales Press, 1966.

Richardson, H. W. "Input-Output and Economic Base Multipliers: Looking Backward and Forward," *Journal of Regional Science*, 25 (1985), 607-661.

Roesler, T. W., F. C. Lamphear, and M. P. Beveridge. *The Economic Impact of Irrigated Agriculture on the Economy of Nebraska.* Lincoln: University of Nebraska, Bureau of Business Research, 1968.

Schaffer, W. and K. Chu. "Nonsurvey Techniques for Constructing Interindustry Models," *Papers, Regional Science Association*, 23 (1969), 83-89.

Stevens, B. H., G. I. Treyz, D. J. Ehrlich, and J. R. Bower. "A New Technique for the Construction of Non-Survey Regional Input-Output Models and Comparisons with Two Survey-Based Models," *International Regional Science Review*, 8 (1983), 271-286.

Vanwynsberghe, D. "An Operational Nonsurvey Technique for Estimating a Coherent Set of Interregional Input-Output Tables," in K. R. Polenske and

J. V. Skolka (eds.) *Advances in Input-Output Analysis.* Cambridge, Mass: Ballinger, 1976, pp. 279-294.

Table 4.1. Survey Based Regional Input-Output Studies

Region	Authors
Kansas	Emerson (1969) updated 1985
Wichita, Kansas	Emerson and Reed (1971)
West Virginia	Miernyk, et al. (1970)
Washington	Bourque and Conway (1977)
Philadelphia	Isard, Langford and Romanoff (1966-68)
Nebraska	Roesler, Lamphear and Beveridge (1968)
St. Louis	Hirsch (1959)
Clinton County, Pennsylvania	Gamble and Raphael (1965)
Sullivan County, Pennsylvania	Gamble (1967)
Kalamazoo, Michigan	Hochwald, Striner and Sonenblum (1960)
New Mexico	Bureau of Business Research, University of New Mexico (1967)
Utah	Bradley, Short and Kolb (1970)

Appendix

Federal Government Censuses Containing State Data

1. U. S. Department of Commerce, Bureau of the Census, Census of Manufactures (every five years)
    - Employment by Industry
    - Value Added by Industry
    - Cost of Materials by Industry
    - New Capital Expenditures by Industry

2. U. S. Department of Commerce, Bureau of the Census, Annual Survey of Manufactures (annual)
    - Same Data Sets as Census of Manufactures

3. U. S. Department of Commerce, Bureau of the Census, Census of Agriculture (every five years). Also has county data.
    - Livestock Sales
    - Crops Harvested and Value of Production

- Farm Production Expenses
- Farm Payroll and Employment
- Market Value of Products Sold

4. U. S. Department of Commerce, Bureau of the Census, County Business Patterns (annual)
    - Employment by Industry

5. U. S. Department of Commerce, Bureau of the Census, Census of Mineral Industries (every five years)
    - Employment and Payroll
    - Value Added
    - Cost of Supplies Used
    - Value of Shipments
    - Capital Expenditures

6. U. S. Department of Commerce, Bureau of the Census, Census of Construction Industries (every five years). Also has MSA data.
    - Employment and Payroll by Industry
    - Value Added by Industry
    - Payments for Materials, Components, and Supplies by Industry
    - Payments for Construction Work Subcontracted to Others by Industry
    - Payments for Machinery, Equipment, and Structures by Industry
    - Capital Expenditures by Industry
    - Construction Receipts by Industry

7. U. S. Department of Commerce, Bureau of the Census, Census of Retail Trade (every five years). Also has MSA and county data.
    - Payroll by Industry
    - Sales by Industry

8. U. S. Department of Commerce, Bureau of the Census, Census of Wholesale Trade (every five years). Also has MSA and county data.
    - Payroll by Industry
    - Sales by Industry

9. U. S. Department of Commerce, Bureau of the Census, Census of Service Industries (every five years). Also has MSA and county data.
    - Payroll by Industry
    - Sales by Industry

Interindustry Impact Project

ID No. ______________________

Major products you produce ____________________

Number of establishments covered by this questionnaire ____________

Purchases

Please allocate your 1985 purchases according to supplying industries and location of suppliers. (See attached list of industries).

Supplying Industries - brief description	Total Purchases ($ or %)	Percent Supplied by Producers in Kansas	Percent Supplied by Producers in Your County
Other Expenditures			
Wages and salaries			
Taxes - Federal			
- State			
- Local			
Depreciation and retained earnings			
TOTAL EXPENDITURES			

Sales

Please allocate your sales among the various business, industrial or government customers of your product. (See attached sheet for a suggestion of market types).

Markets - brief description	Total Purchases ($ or %)	Percent of Sales Made in Kansas	Percent of Sales Made in Your County
Household Consumers			
Kansas State Government			
Kansas Local Government			
Federal Government, non-defense			
Federal Government, defense			
TOTAL RECEIPTS			

Interindustry Impact Project
Sector Listings

FARMING
1. Corn
2. Sorghum
3. Wheat
4. Other Grains
5. Soybeans
6. Hay
7. Dairy Products
8. Poultry and Poultry Products
9. Cattle
10. Hogs
11. Other Agricultural Products
12. Agricultural Services

MINING
13. Crude Oil and Natural Gas
14. Oil and Gas Field Services
15. Nonmetallic Mining
16. Other Mining

CONSTRUCTION
17. Maintenance and Repair
18. Building Construction
19. Heavy Construction
20. Special Trade Construction

MANUFACTURING
21. Meat Products
22. Dairy Products
23. Grain Mill Products
24. Other Food and Kindred Products
25. Apparel
26. Paper and Allied Products
27. Printing and Publishing
28. Industrial Chemicals
29. Agricultural Chemicals
30. Other Chemicals
31. Petroleum and Coal Products
32. Rubber and Plastics
33. Cement and Concrete
34. Other Stone and Clay
35. Primary Metals
36. Fabricated Metals
37. Other Fabricated Metal Products
38. Farm Machinery
39. Construction Machinery
40. Food Products Machinery
41. Electrical Machinery
42. Other Machinery
43. Motor Vehicles
44. Aerospace
45. Trailer Coaches
46. Other Transportation Equipment
47. Other Manufacturing

TRANSPORTATION
48. Railroad Transportation
49. Motor Freight
50. Other Transportation

UTILITIES
51. Communications
52. Electric Gas and Sanitary Services

WHOLESALE
53. Groceries
54. Farm Products
55. Machinery and Equipment
56. Other Wholesale Trade

RETAIL
57. Farm Equipment Dealers
58. Gasoline Service Stations
59. Eating and Drinking
60. Other Retail Trade

F.I.R.E
61. Banking
62. Other Finance
63. Insurance and Real Estate

SERVICES
64. Lodging Services
65. Personal Services
66. Business Services
67. Medical and Health Services
68. Other Services
69. Education

5

Nonsurvey Approach to I/O Modeling

Curtis H. Braschler and Gary T. Devino

Introduction

Since the development of the input-output model by W. Leontief, the model has had substantial theoretical and empirical appeal as a tool for national and regional economic analysis. Given the necessary data, the model can be used for both forecasting and economic impact analysis.

In its most basic formulation, the model requires large amounts of primary economic data. Data requirements for an I/O analysis can best be illustrated by considering a mechanical trading paradigm for a specific geographical entity such as nation, state, or substate region. The mechanics of geographical exchange involves trading between firms, firms and consumers, and firms and all levels of government (local, state and federal). In addition, a particular region will also export to other regions and purchase goods and services from other regional entities. These are imports and exports for a region and involve the exchange of goods and services between regions.

In essence, economic activity at a point in time involves a flow of money for goods and services between business firms, consumers and business firms, business firms and local, state and federal governments. It is the monetary transactions between these units of decision with an opposite flow of goods and services over a specified time period (usually one year) and a geographical entity which provides the basic data for the construction of an I/O model. These data may be obtained by surveying either a total population of firms within each industry category or of a representative sample of firms in an industry that can be carefully selected to represent that industry.

Data requirements expand as the square of the number of industry categories are increased. For example, a 10 intermediate sector model would require 10^2

or 100 elements of data. Obviously, a detailed survey of each industry's sales and purchase for construction of an I/O model has become increasingly expensive in both time and money for state and substate analysis. Increasing costs of survey collected data have been a major impetus for researchers in regional analysis to seek lower cost methods for the acquisition of data required for detailed regional analysis.

Nonsurvey Methods

Nonsurvey Input/Output models that utilize secondary data rather than the much costlier primary data obtained from direct surveys have the distinct advantage of being less costly than survey approaches. However, both survey and nonsurvey methods require a fairly high level of understanding of the mathematics of the model as well as the economic interpretations of the model. Public or private decision makers will usually need technical assistance for developing and interpreting results.

The estimation of total sales from secondary sources raises concern among regional analysts about the accuracy of nonsurvey approaches. Persons using nonsurvey approaches need to be aware of possible sampling bias and to work to minimize its effects.

Nonsurvey Input/Output Models

In this method model requirements are developed from secondary data sources, primarily from federal government and some private sources of regional economic data. A nonsurvey study will typically start with a survey based I/O model of another region, usually the national model, and try to adapt it to the region of interest. The adaptation requires estimation of the total sales to other industries and to final demand in the region by sector of industry.

There are three major nonsurvey I/O analysis procedures: (1) The Locations Quotients, (2) Supply-Demand Pool, and (3) Regional Purchase Coefficients.[1] Most of the nonsurvey studies that have been completed in recent years used one or more of these methods for adjustments of national coefficients or some other previously developed regional I/O data as a data source. In this discussion we will assume that readers will have general understanding of the various components of I/O analysis.[2]

The three procedures discussed in this paper require an accurate measurement of the total local output of each industry included in the study in monetary units.

These different procedures share a common characteristic in that these modify the national I/O input coefficient matrix. These industries are assumed to have similar (the same) production functions as the national technology matrix. Some industries will produce enough for the local region or substate region and others will not likely produce all of the goods necessary for internal production of some of the goods utilized in the region. Some industries in a given region will produce more than can be utilized locally. These industries will export output from the region to other regions or will export to other countries. They need no modification for local application. Other industries in the region will produce less of a good than is needed for intermediate industries and final demand requirements for the region. Such deficiencies will be met by imports from outside the region.

First, total outputs by all industries must be determined from secondary sources. These outputs include sales to other processing industries as well as sales to final demand components of the regional economy. In most situations gross industry output will not be available directly from secondary sources. Analysts need to use earnings, employment, income or some other proxy for total individual industry output. The determination of estimates of regional output will be designated x_i. This value will be assumed to be the actual industry output. There are some possibilities for error. This needs to be recognized and a comparison with Census data or other secondary data sources should be made, if possible.

The next step involves the determination of an estimate of required industry output given by the structure of the regional economy. It is in the determination of this "required" local output that the three major methods differ in terms of basic assumptions. It should be noted that a comparison of the actual output x_i with the "required" local output is what determines whether national input coefficients need to be modified in a particular industry, x_i, assuming local production methods are the same as national input coefficients.

It is in the case where local output of x_i is not sufficient for local demand that a modification of national coefficients are required for an analysis of the regional economy. It is in the methods used to determine local demand where differences arise in the choice of reduction techniques.

Location Quotients

The Simple Location Quotients procedure computes a quantity s_i that is designated as the production of x_i assuming that the regional economy has the same industry mix as the national economy.

$$(1) \quad s_i = X_i \left(\frac{x}{X}\right)$$

where: s_i = is the estimated production of x_i in the region
X_i = the national output of x_i

x = the output of all industries in the region of study
X = total output in all industries of the national economy

While the Location Quotient measures the required output of x_i that will satisfy the demands of the processing sectors, it does not directly yield regional values for final demand vectors. If local estimates of final demand requirements are not available, these values must be estimated by determining the regional share of national final demand. It is usually assumed that the regions share is in proportion to national final demand.

Supply-Demand Pool

The Supply-Demand pool approach is based on satisfying local requirements and final demand from regional sources first. The quantity

$$(2) \quad d_i = \Sigma_j x_{ij} + \Sigma_t f_{it}$$

where: d_i = the total regional demand for the output of industry i

$$(3) \quad x_{ij} = A_{ij} x_j$$

where: A_{ij} = the national I/O coefficient for industry i
x_j = the estimated regional output of industry x_i
f_{it} = is the final demand estimates of regional final demand for industry i.

By assuming regional final demand is the regional share of national final demand, i.e., regional final demand is in the same proportion as is the national final demand. The quantity d_i is an estimate of the total requirements in the region to satisfy local industry and final demand and is the Supply-Demand pool estimate of local requirements. It is the Supply-Demand conceptual equivalent of the value s_i of local use of industry x_i determined by Location Quotients. The quantity d_i is perhaps more difficult to interpret conceptually than is the quantity s_i. The quantity d_i accounts for the difference in the local mix of industries in the region, the quantity s_i does not. Nevertheless, the two values are conceptually the same but are quite different empirically because of the assumptions made in calculating the estimates of the quantity x_i when quantity x_i is determined or estimated from local employment, earnings or income.

The determination of local industries that can use the national coefficients are those that must be modified for local use are determined by the ratio x_i/s_i in the LQ_i case and by the ratio x_i/d_i when the pool approach is used for reduction of national I/O coefficients. If the ratio x_i/s_i is numerically $\geq$ to 1, then the local economy is producing as much or more of x_i than is needed for intermediate sectors and final demand. The excess output is assumed to be exported. In this case the national A_{ij} coefficients are assumed to be correct for the local economy. If the ratio of x_i/s_i is < 1, the national A_{ij} must be reduced by the ratio of x_i/s_i, i.e., $a_{ij} = x_i/s_i \cdot A_{ij}$. Similar interpretations for reducing A_{ij} by the $a_{ij} = A_{ij} \cdot x_i/d_i$ when the pool procedure is used for reduction of A_{ij} to regional input a_{ij}.

Regional Purchase Coefficients

The other major procedure used to modify a national I/O matrix to a smaller region is called the Region Purchase Coefficient or RPC_i. The conceptual definition is:

$$(4)\quad RPC_i^R = \frac{(X_i^R - E_i^R)}{(X_i^R - E_i^R + M_i^R)}$$

where: X_i^R = is the total regional output of industry i,
E_i^R = the gross amount of the output of industry i that is exported from the region
M_i^R = the gross amount of industry i that is imported.

The conceptual definition is relatively easy to understand. The numerator indicates the amount of the production of X_i that is available to purchasers in region R and the denominator is the total amount of X_i that is available in region R, either produced locally or imported.

Although the conceptual definition is relatively easy to understand, the operational definition is very difficult to determine because of lack of local or regional data estimates of gross exports and gross imports of each industry. Two major developers and users of empirical measures of the Regional Purchase Coefficient have been the Regional Economic Models, Inc. (or REMI), developed by Stevens, Treyz, Ehrlich and Bower and the USDA Forestry Service by Alward, et al, which is utilized in the current version of the IMPLAN I/O model. These researchers utilize data on commodity production, employment, proportional regional to U.S. levels of employee compensation, and land area to empirically estimate parameters which determine regional RPC values.

It should be noted that the quantities s_i, d_i, and RPC_i are all different empirical methods of determining the industries in a region, R, which are producing enough local output to satisfy intermediate and final demand requirements for industry X_i. These industries can use national I/O input coefficients and are net export industries. In those industries where the ratios of x_i/s_i or x_i/d_i are less than one it is indicated that total local output x_i is not sufficient to meet local requirements. The national A_{ij} coefficients must be reduced by that of the ratio based on a quantity that is less than one.

Comparison of Survey and Nonsurvey I/O Model

Given the number of variations in these three basic methods of data reduction it is not surprising to find strong disagreement about the use of nonsurvey techniques among regional analysts. Schaeffer and Chu in their seminal study published in 1969 examined five different nonsurvey methods of reducing national I/O coefficients to the regional level including several variations of the Location Quotients approach. The study also used a survey analysis which had been done previously for comparison with the nonsurvey procedure. These investigators concluded that these nonsurvey methods might be useful supplements to survey studies, although there was no acceptable substitute for a good survey-based study.

In contrast, Cartwright and Beemiller discussing the Rims II I/O model (Regional Input-Output Modeling System) concluded that the Location Quotient procedure resulted in only 5 to 10 percent larger industry specific multipliers than expected with direct surveys.

Round compared several different variations of the Location Quotients approach to determine estimates of regional tables. He found no significant difference in the I/O multipliers developed with the different methods for analysis of the alternative survey procedures. Miller and Blair arrive at somewhat different conclusions. Their evaluations suggested that the more sophisticated versions of Location Quotients or pool procedures offer enough improvement in accuracy to justify their use.

Miller and Blair also evaluated the RPC approach and are basically noncommittal regarding the use of this method. They conclude that further refinement or more secondary data on regional imports and exports would greatly enhance the RPC procedure's accuracy for projection and impact analysis. Miller and Blair also concluded that the RAS partial survey procedures may become a good choice for regional reduction of I/O tables. It does use row and column vector data which may not be available other than by direct survey for a region.

Procter, Braschler and Kuehn used nonsurvey methods for an I/O study of Missouri. Simple Location Quotients and the Pool procedure were compared with the results from a partial survey study by Harmston and also an economic base analysis developed earlier by Braschler and Kuehn. Type two multipliers were compared. As was expected from the earlier studies by Schaeffer and Chu and Round, multipliers developed from nonsurvey procedures were higher for nonsurvey methods than partial survey methods used by Harmston.

Summary

In this paper we have reviewed three major procedures i.e., Simple Location Quotients, Supply-Demand Pool and Regional Purchase Coefficient Configurations for reducing national I/O coefficients for regional applications. Limited time and space precluded a review of all studies using these procedures. We are aware that some variation of these procedures have been used and would appear to be refinements of the three basic procedures. These procedures are essentially the same in terms of their conceptual definitions. All assume that production functions for each industry are for practical purposes the same as at the national level. This assumption is necessary for application of any of the three procedures. A crucial requirement for regional applications is an estimation of local industry output for all industries defined in the region. These estimates usually have been made from secondary sources and include BEA and county business pattern data from the U.S. Department of Commerce, employment, income and earnings and data by industry. These data have certain limitations in terms of SIC codes available and suppressions of certain local values when individual firms might be identified. Methods have been developed to estimate these values but errors in estimation can occur. Nevertheless industry output estimates must be assumed to be the best estimate of actual output.

It is then necessary to determine a regional output that will supply intermediate and final demand requirements for the region in each industry. It is in this area where a determination of required output will differ in terms of the use of one of the three different approaches. The required output is compared to the actual industry regional output. If actual industry output is larger than required, the national I/O direct input coefficients are assumed applicable to the region. When actual output is less than the required output, then it is assumed that the deficiency will be imported. The national coefficients must be reduced in terms of the proportion that is imported for all industries requiring imports to meet local requirements.

It appears that given cost considerations, the direct survey approach for local impact analysis and forecasting will be used less in the future than lower cost nonsurvey I/O and economic base analysis. The three major methods of nonsurvey I/O modeling will likely be refined with additional use. The issue of accuracy of these methods and estimation procedures will also continue to concern regional analysts.

Notes

[1] In addition to the three nonsurvey methods, there is a partial survey method known as the RAS procedure. Because this procedure is still in the development stage it will not be reviewed in this paper.

[2] For the reader not familiar with the general framework of an I/O model there are numerous published books and research bulletins that give a thorough review of the mathematics as well as economic interpretation of I/O models. McKean, et al., develops a particularly excellent and easy to understand version. More detail can be obtained from a book published by Miller and Blair in 1985. This book can probably be considered "state of the art" even though it has been available for three or more years. Such deficients will be met by imports from outside the region.

References

Alward, G.S., H.C. Davis, K.A. Despotakis, and E.M. Lofting. 1985. "Regional Non-Survey Input-Output Analysis with IMPLAN." Paper presented at the Southern Regional Science Association Conference, Washington, DC.

Braschler, Curtis. 1972. "A Comparison of Least-Squares Estimates of Regional Employment Multipliers with Other Methods." *Journal of Regional Science,* Vol. 12, No. 3, pp. 457-68.

Braschler, C. and J.A. Kuehn. 1975. "Industry Sectors and the Export Base Determination of Nonmetropolitan Employment Change in Four Midwestern States." *Review of Regional Studies* 5(3): 82-89.

Cartwright, J.V., R.M. Beemiller, and R.D. Gustley. 1981. *Regional Input-Output Modeling System.* U.S. Department of Commerce, Bureau of Economic Analysis.

Harmston, Floyd K., Vamon Rao, Jasbir S. Jaswal, and Wayne S. Chow. *Intersectoral Analysis of the Missouri Economics 1958, 1963, 1967, 1972.* Vol. 1: *Intersectoral Flow of Goods and Services in Current and Constant Dollars.*

Kuehn, John A., Curtis Braschler, and John A. Croll. 1982. "Economic Base Multipliers and Community Growth." DM3006, Cooperative Extension Service, University of Missouri-Columbia, August.

Leontief, Wassily. *Input-Output Economics.* 1966. New York: Oxford University Press.

Leontief, Wassily. *1951. The Structure of American Economy, 1919-1939.* New York University Press.

McKean, J.R., W.P. Spencer, and W.D. Winger. 1988. "Projecting Economic Impacts of New Industry in Colorado with IMS Micro-computer Software and USDA IMPLAN Data Synthesizer." Technical Bulletin LTB88-1, Department of Agricultural and Resource Economics, Colorado State University.

Miller, Ronald E. and Peter D. Blair. *1985. Input-Output Analysis: Foundation and Extensions.* Prentice-Hall, Inc., Englewood Cliffs, New Jersey.

Procter, Michael H. 1982. "Non-survey Input-Output Models for Missouri: Development and Evaluation." Ph.D. Dissertation, Department of Agricultural Economics, University of Missouri-Columbia, August.

Round, Jeffery I. 1978. "An Interregional Input-Output Approach to the Evaluation of Nonsurvey Methods." *Journal of Regional Science* Vol. 18, No. 2, pp. 179-94.

Schaeffer, William and Kong Chu. 1969. "Nonsurvey Techniques for Constructing Regional Interindustry Models." Papers, *Regional Science Association* 23: 83-101.

Stevens, Benjamin H., George K. Treyz, David J. Ehrlich, and James R. Bower. 1983. "A New Technique for Construction of Nonsurvey Regional Input-Output Models and Comparison with Two Survey-Based." *Regional Science Review* 8, No. 3 (December): 271-83.

6

Developing or Selecting a Regional Input-Output Model

Sharon M. Brucker and Steven E. Hastings

Introduction

Over the past decades regional economic researchers and planners in the academic, business, and government sectors have become increasingly involved in the analysis of economic growth and/or decline in rural communities and regions. Furthermore, funding for development projects of many sorts has frequently been conditional on the provision of estimates of economic impacts of the project. Together, the concerns for regional development and funding requirements have led to increased use of analytic models to assess economic impacts. During this same period, the nature of such analysis has undergone major changes. One of the most popular methods, regional input-output analysis has benefitted from many improvements in estimation techniques which have reduced the limitations of the model. Furthermore, the costs in both time and money required to estimate a model have been reduced by vast changes in computing technology.

The combination of these lowered costs and increases in derived demand has led both to the increased popularity of input-output with regional analysts and to an expanded market supply of input-output "tools" (software) and "ready-made," often commercially available, modeling systems.

Thus, the choices facing users of input-output analysis are greater both with respect to sources of models and with respect to techniques and tools available to estimate their own models. The purpose of this chapter is to briefly present the array of choices facing a user and describe the factors that would be appropriate to include in making a choice.

In discussing input-output analysis, it is helpful to divide the work into three stages: planning, construction and application. One of the results of the changes in the market for input-output models has been a reallocation of an input-output analyst's resources among these three phases. In the past, a large amount of the time and money was spent in the construction phase of the model building (Figure 1). Because of the lack of options for acquiring a model, the planning stage largely entailed lining up funding sources for the extensive personnel, surveying and computing costs of estimating the model. Because the funding needs were so substantial, the selection of a region, objectives, and sectors were usually dictated by the source of funds. The enormous task of identifying through a survey the transactions of all the sectors of a regional economy made the construction stage very lengthy. In many instances, when the model was completed after several years of work, there was frequently little time, energy or funding left to make use of the wide variety of information available in the model. It was frequently disseminated through cumbersome written reports. Even in the settings where input-output models were widely used for varied applications,[1] the availability of the information was often dictated by a potential user's ability to access a mainframe computer and the time and hardware necessary for such access.

In contrast, modelers today face many options of models, modeling tools and procedures. In order to acquire the best model for their purposes, the potential user must spend a fairly large amount of time defining needs, assessing resources, and matching them with the model with the characteristics which best meet the needs subject to the resource availability. Therefore, the time spent in planning is greater since it requires more information and involves a greater number of choices. Conversely, the improvements in computer technology and the market's provision of new tools assure that the time necessary to build or acquire a model today will be less than the time spent to acquire an identical model 15 years ago. As indicated in Figure 1, even the most labor and funds intensive type of model (a full survey model) will take considerable less time to build today with faster computers, easier data entry, and input-output estimation software available for desktop personal computers.

Resources saved in the model acquisition stage can be utilized in the use phase to improve the accuracy of the regional input-output model. Thus, today's users spend a larger percent of their time in the use of the model. During the use phase a modeler can do three types of things to improve the accuracy of impact predictions: 1) improve final demand change identification, 2) analyze disaggregated impacts (to identify development, employment target opportunities, and assess capacity constraints, reality checks), and 3) in flexible models, adjust for technology, trading patterns, or price effects unique to a region or situation. Furthermore, the freed resources can be expanding the applications of the model to such areas as pollution impacts, fiscal impacts, community resource impacts and other useful analyses.

This brief discussion of the allocation of resources among the stages of input-output analysis highlights the fact that there are many choices in each stage that a modeler must make. However, the scope of this chapter will be limited to the options for acquiring a model and the appropriate factors to consider when making that choice.

It may be important to note at the outset that it is unlikely that there is one best model for all users. The variety of models that exist reflects the diversity of needs that exists. In Table 6.1, the models are classified on the basis of the degree of the involvement of the user in their construction.

The process of choosing a regional input-output model entails the matching of the model requirements to the resources available and matching the desired output (based on types of use and objectives) to the output the model can provide. This matching exercise can be facilitated by identifying the objectives from the suggested categories in Table 6.2. It is equally important to take inventory of the resources available similar to the categories listed in Table 6.2. There are studies which allow the decision maker to categorize the models according to resource requirements and output criteria in Table 6.3 to determine a suitable match. See Table 6.1, footnote 1.

Several examples may clarify how this choice will differ for different resource and objectives situations. User A wants to analyze the impacts of an additional branch of an existing industry moving into his region, has very little computer knowledge or understanding of input-output methodology and has very limited financial resources. He may want to limit his choices to buying RIMS II for his county or getting IMPLAN.

However, User B, is a knowledgeable input-output practitioner, has a frequently-used PC on her desk and has access to and familiarity with several state-wide data sets as well as standard secondary data sets. She wants to analyze the impacts on the state of a new industry locating in the state as well as the impacts of several recent zoning changes on state revenue. She will be able to have a more accurate and flexible model if she acquires IMPLAN or ADOTMATR and builds her own transactions table. User C has no interest in learning input-output methodology and doesn't like computers, but has a major grant to determine the impact of increased tourism on the local resort economy. User C has more money than other resources and needs the expertise of a consultant. User C would be wise to hire Schaffer, Stevens or Lamphear to determine the impacts for them. For User D, a major research effort is underway to understand the economy of a region and several research professionals are committed to a survey effort to describe their region's transactions; this group would be well-served to acquire ADOTMATR as a computer shell within which to construct and then manipulate their input-output tables.

Finally, a state development office has an eager college graduate working for them who is willing to read Chapters 4 and 5 of Miller and Blair. This individual is familiar with using a spreadsheet and wants to do frequent and varied analyses of impacts of structural and policy changes on the region. In this case, the user would be wise to get an IMPLAN, RSRI or RIMS II model for their region (the decision would be based on cost and on need for specific special impacts available on the higher priced RSRI model) and take the time to learn how to use it accurately. From these five simplified examples, it is obvious that various combinations of resources and objectives can lead different users to different choices of the model that is best for them.

Other things being equal, users will want to identify the most accurate model. It is generally assumed that a survey-based model would be the most accurate; however, the decision process outlined above leads most users to determine that the time and monetary costs of collecting primary data is prohibitive. Therefore, the question becomes which of the nonsurvey based methods or models is most accurate. Research on this question usually compares nonsurvey based models to survey based estimates using closeness to survey results as a measure of accuracy. In a recent study,[3] no consistent pattern of one model being more similar to the survey-based model than any other was found. The only pattern identified was that the estimates of total region wide production changes and income impacts were closer to the survey model than were the employment impacts. This was also true if the several models were compared to the mean of their own estimates. In other words the several nonsurvey models, regardless of regionalizing procedure or other differences in methodology, were most in agreement as to production and income effects, but were widely divergent in estimates of employment effects. This would suggest that improvements in accuracy are most likely to come from better estimates of direct employment effects, based on primary data or at least adjustments to secondary data where actual employment patterns are known to be unique to the region or to a specific situation.

Since most users find the costs of a full survey based model prohibitive and since most users can acquire a nonsurvey based model at a relatively low cost of time and money, modelers encourage users to spend some time and money in the use of the model they build or buy. There are three types of efforts that are suggested. First, users will find many of the models flexible enough to accommodate additional data supplied by users. Collecting and incorporating such data into the direct requirements and labor coefficients (as discussed above) of a model customize the table to better depict the region. Such a "hybrid" model has some of the accurate detail of a survey model and the ease and lower cost of nonsurvey models. Second, users are urged to spend some time identifying the initial change that is expected to take place in the economy and which is being analyzed. The single most important part of measuring the total region wide impact of a given change is the magnitude and nature of the initial change itself. If the initial change is incorrectly quantified or attributed to the wrong sector, then the multiplier type of analysis will be multiplying the error.

Therefore, a rather careful process to assure the initial change is correct is essential to an accurate estimate of the impacts. Such a process needs to involve interaction with local planners and regional industrial experts as well as a thorough understanding of the assumptions of the model. Third, users are encouraged to become familiar with the extensive disaggregated information provided by the model. Using this information to predict changes in the economy can give policy makers more avenues for action than a limited single multiplier analysis.

In summary, the best choice of a regional input-output model will depend on the "fit" of the model to the resources and objectives of the user. Meaningful analysis will depend much more on the care taken in using any model than on the choice of the model to use. The mechanics of the models assure that the multiple impact of a *given change* will be quite similar regardless of the model chosen. Relatively speaking, the magnitude of the difference in total regional impacts will depend much more on the size and nature of the initial change to be evaluated than on the multiplier. Therefore, the advent of "ready-made" models may have contributed to regional input-output analysis most by freeing up resources that can be spent in better identifying the initial change and by being flexible and easy-to-use so that they facilitate and encourage the collection of data specific to the region to be studied.

Notes

[1] In research institutions such as West Virginia University, Washington State University, Kansas State University, the University of Pennsylvania, Georgia Institute of Technology, the Bureau of Economic Analysis, and the Regional Science Research Institute, where professors and their students and staff were dedicated to ongoing models, models were constructed by craftsmen and used extensively for creative and appropriate applications.

[2] See Brucker, Hastings and Latham (1987).

[3] See Brucker, Hastings, and Latham (1988).

References

Brucker, S. M., S. E. Hastings, and W. R. Latham. "Regional Input-Output Analysis: A Comparison of Five Ready-Made Model Systems." *Review of Regional Studies*. Spring, 1987.

Brucker, S. M., S. E. Hastings, and W. R. Latham. "Analysis of Estimated Impacts from Five Regional Input-Output Modeling Systems." *International Regional Science Review*. Vol. 12, No. 2.

Table 6.1. Ways to Acquire a Model[1]

Buy Consultant Services	Buy and Use "Ready-Made" Model	Construct Own With Shell	Construct Own Write Program
Schaffer	RIMS II	ADOTMATR	Use mainframe
Stevens (RSRI)	RSRI	IO/EAM	Use PC
Lamphear	IMPLAN	IMPLAN	
Treyz (REMI)	REMI		
Other			

[1] See Brucker, Hastings and Latham (1987) for additional information about these models.

Table 6.2. Categories of User Needs and User Resources

User Needs	User Resources
Planned Uses	Human Capital
Description	Computer skill
Growth Targets	Familiarity with I/O
Impacts	Number of people
Region wide	People's time
Distribution by sector	
Distribution by income	Physical Capital
	Computer hardware
Dissemination	
Users of Model	Money
Centralized analysis	
Multiple decentralized users	Data Availability
Method of Model Access	
Written Report	
Computer Use	
Phone Consultation	
Nature of Region	
Political Unit	
State, County or Municipality	
Geographic/Economic Unit	
SMSA or market Area	
Multiple or Changing Units of Analysis	
Number and Definition of Sectors	
Time Horizons	
Construction Phase	
Time until completion	
Use Phase	
Frequency of use (once, monthly, annually)	
Time period of prediction	

Table 6.3. Categories of Resource Requirements by Models and Model Characteristics

Resource Requirements	Model Characteristics
Human Capital	Potential Uses
Computer skill	Types of multipliers
I-O expertise	Round-by-round impacts
Time required	Transactions table
Construction	Disaggregated impacts
Use	Full model or multipliers only
	Additional types of information
Physical Capital	Demographic impacts
Computer hardware	Environmental impacts
For construction	National coefficients
For use	Bridge tables
	Trade margins
Monetary Costs	Import margins
Data Requirements	Format for Dissemination
	Printed output/one time
	PC/software/user runs/centralized
	PC/software/user friendly/novice
	Flexibility
	Sector definition/aggreg.
	Region definition
	Incorporate user data
	Multiple uses
	User-defined change vector
	Different regions
	Methodology
	Regionalizing Procedures
	Open/closed model
	Sector definitions

Figure 1

Time and Resources Spent on
Input-Output Analysis

15 YEARS AGO	TODAY
PLANNING Choose Region Define Sectors Get Funding Personnel Computer	PLANNING More Models Varied Applications Levels Disaggreg. Regions Sectors Less Funding Required
MODEL BUILDING Design Questionnaire Identify Sample Collect Data Code and Enter Data Write Computer Program Interpret Results Report Results -Written Re-run program for each analysis	MODEL ACQUISITION Buy Build Less time data entry Software tools Instantaneous Additional Runs Adjustments
USE Look up multiplier	USE Improve Final Demand I.D. More Accessible for "What if" Have Disaggregated Impacts Have Flexibility to Adjust Reality Checks Change Technical Coefficients Acknowledge Capacity Constraints Update Frequently Expert Input

7

Regional Economic Impact of the Conservation Reserve Program: An Application of Input-Output Analysis

F. Larry Leistritz, Timothy L. Mortensen, Randal C. Coon, Jay A. Leitch, and Brenda L. Ekstrom

Introduction

The Conservation Reserve Program (CRP) was authorized by the 1985 Food Security Act (Public Law 99-198) and was passed at a time of heightened concern for environmental quality. Its main objective is to take highly erodible land out of production, thereby reducing wind and water erosion, protecting long-term food-producing capability, reducing sedimentation, improving water quality, creating wildlife habitat, curbing excess production, and providing income support for farmers.

Landowners who wish to participate in CRP must agree to implement a conservation plan that provides for permanent vegetative cover on the land for ten years. In return, the federal government pays the landowner an annual contract payment determined by a bidding process. Land entered must be classified as "highly erodible" by USDA Soil Conservation Service personnel, and no more than 25 percent of an individual county's total cropland may be entered into CRP without USDA approval.

Because the present program has objectives similar to those of the Soil Bank Program of the late 1950s, concern has been expressed in areas with high concentrations of eligible land regarding possible economic impacts of the program.[1] Potential impacts that have been identified include those arising from (1) reduction in use of agricultural inputs such as fuel, fertilizer, and chemicals;

(2) reduction in the use of farm labor and machinery; and (3) long-term changes in land use if CRP land is not returned to crop production at the end of the contract period. The analysis reported here was undertaken to estimate the short-run economic impacts of the CRP program in North Dakota (i.e., those arising from reductions in use of agricultural inputs).

Procedures

The study had two major phases. First, a statewide survey of CRP participants was conducted to determine selected characteristics of those individuals and their enrolled land that would be important for subsequent impact estimation. These characteristics included land attributes (such as comparison of costs and returns and soil productivity to those of non-CRP land in the area, comparison of CRP payments to local cash rents, cover option chosen, and cost of cover establishment) and landowner characteristics (such as age, residency, level of farm income, and use of CRP payments). A questionnaire was mailed to nearly 3,000 randomly selected landowners in North Dakota (approximately 40 percent of all participants) in early March 1988. Follow-up mailings resulted in 1,289 useable surveys for a response rate of 44 percent. Response rates were quite similar for each of the state's five pool groups (see Figure 1).

Key survey results were tabulated, then a regional input-output model, previously developed from primary data and consisting of 17 sectors, was used to estimate the indirect effects of the CRP program for each of the state's five pool groups. (For a detailed description of the model, see Coon et al. 1985.) An important prerequisite to providing these indirect effects was estimating the direct effects of program participation on farm expenditures and income. Sectors expected to experience direct effects were (1) the retail trade sector; (2) finance, insurance, and real estate; (3) business and personal services; and (4) the household sector (see Table 7.1). The procedures used to estimate these changes in expenditures are summarized in Figure 2. Three main sources of data were used to estimate expenditure changes: (1) county CRP survey data (Mortensen et al. 1988), (2) North Dakota agricultural statistics, and (3) county data from the state Agricultural Stabilization and Conservation Service (ASCS)[2]. Initially compiled on a county-by-county basis, the resulting estimates fall into three main categories: (1) reduced input expenditures, (2) reduced federal commodity payments, and (3) increased CRP contract payments and upkeep costs. (For a more detailed discussion of data sources and estimation procedures, see Mortensen et al. 1989.)

After the change in business activity resulting from the CRP program had been estimated for each sector, the resulting change in employment was estimated based on historic relationships between employment and gross business volume in each sector.

Results

CRP participants generally felt their CRP land was less productive than other land in the area and that input costs were slightly higher (Table 7.2). (Unless otherwise noted, the values shown are the mean for all survey respondents.) CRP contract payments were felt to be 6.7 percent higher, on average, than prevailing cash rental rates in the area. The initial cost of establishing CRP cover averaged $37.20 per acre with more than 42.4 percent of responses falling between $30 and $40. Annual maintenance costs averaged $6.92, while annual contract payments averaged $36.98. More than 60 percent of all contracts had annual payments of $30 to $40.

The average age of the CRP landowners was 57 years, and 90 percent lived in North Dakota (Table 7.2). About 73 percent of the respondents had farmed either full- or part-time in 1987. For the farmers, the average gross farm income for 1987 was just over $94,000, or about 20 percent less than that reported for that year by a statewide longitudinal farm panel (Leistritz et al. 1989). The average net cash farm income of $16,259 was about 22 percent less than that for the farm panel. For 41 percent of these producers, their CRP income exceeded their net cash farm income, and about 21 percent said that the program enabled them to continue farming.

Reduced direct expenditures caused by taking CRP land out of production total $55 million for the state with nearly 62 percent impacting the retail sector (Table 7.3). Pool groups two, four, and five have the highest net impact at about $12 million each. The household sector is positively affected in pool groups one, two, and three primarily because the CRP rental payments exceeded the farm income and government program payments that were foregone.

The direct effects were applied to the input-output model to estimate the total impact of the CRP program. Table 7.4 summarizes baseline business activity (i.e., estimated gross business volume or gross receipts of the respective sectors for the period 1980-87); the changes in business activity associated with CRP-related reductions in expenditures; increases in household incomes; and the net effect of the CRP program on business activity in each sector. The $55 million in direct effects resulting from the CRP result in about $141 million in reduced business activity for the state--an overall multiplier of 2.56. This total is spread among 13 sectors of the state's economy with the retail sector absorbing the greatest impact--about 40 percent of the state total.[3]

Among the county groupings, pool group five had the largest absolute impact from the CRP, reflecting the more intensive nature of agriculture in eastern North Dakota (Table 7.5). Pool group four, on the other hand, had the greatest percentage impact. In no case, however, did the CRP impact exceed 1 percent of the area's baseline business volume. Employment effects of CRP were distributed somewhat differently than effects on business volume; pool group two had the largest total impact. Although the total CRP-related potential employment reduction was estimated to be only 2,416 jobs statewide, or about 0.77 percent of average annual employment in 1987, it should be noted that much of this employment loss may be concentrated in the state's most agriculturally dependent rural areas--areas already hard-hit by reductions in retail trade volume and employment stemming from the depressed state of the agricultural economy.

Conclusions and Implications

The results of this analysis of the impact of the Conservation Reserve Program on the North Dakota economy indicate that impacts of the program to date have been modest at the state and substate regional levels; total business activity was reduced by only 0.54 percent for the state and 0.91 percent for the most substantially affected region. However, it should be noted that the impacts are not distributed uniformly among sectors or communities. Rather, the retail sector accounted for more than 40 percent of the total impact of the program. Further, within the retail sector, businesses that rely on farm supplies or machinery for much of their volume are likely to be affected much more than others. Similarly, because the CRP enrollment varies substantially among counties, those with higher percentages of their land enrolled will obviously experience greater impacts. In North Dakota, five counties had more than 10 percent of their land enrolled through the fifth sign-up (July 1987), and in one county about 22 percent was enrolled. Finally, because substantial acreages have been enrolled in the program in subsequent sign-ups (statewide about 800,000 more acres were added in the sixth and seventh enrollments), the effects of the fully implemented CRP program will be greater than those shown here.

In addition to the negative effects resulting from initial reductions in agricultural activities, the program has a number of positive aspects. A short-run impact has been to sharply increase the demand for grass seed used in establishing vegetative cover. Other, longer-run effects could stem from achievement of the program's conservation objectives, particularly if much of the land remains in noncrop uses after the contracts expire. Estimating possible economic consequences of such effects as reduced soil erosion, increased water quality, and enhanced wildlife habitat was beyond the scope of this study. Such impacts should be addressed in future analyses, however, and input-output analysis would be a very appropriate tool for quantifying some of these effects.

Notes

[1] For a discussion of some impacts of the Soil Bank Program, see Taylor et al. 1961, Barr et al. 1962, and Brown and Weisberger 1958.

[2] Impacts of the CRP were analyzed using 1987 data on farm prices and costs and CRP acres through the fifth sign-up due to availability of data and the abnormal nature of the 1988 drought. It should be recognized, however, that not all acres that were enrolled through July 1987 were taken out of production that year.

[3] Although much of the impact is observed in the retail sector, only the retail margin technically stays in the county. A margining procedure identifies the share at the producer level of each of the expenditure categories.

References

Barr, Wallace, Richard R. Newberg, and Mervin G. Smith. 1962. *Major Economic Impact of the Conservation Reserve on Ohio Agriculture and Rural Communities*, Research Bulletin 904, Wooster: Ohio Agricultural Experiment Station.

Brown, William G., and Pius Weisberger. 1958. "An Appraisal of the Soil Bank Program in the Wheat Summer Fallow Area of Oregon," *Journal of Farm Economics*, XL, No.1: 142-48.

Coon, Randal C., F. Larry Leistritz, Thor A. Hertsgaard, and Arlen Leholm. 1985. *The North Dakota Input-Output Model: A Tool for Analyzing Economic Linkages*. Fargo: North Dakota State University, Department of Agricultural Economics.

Leistritz, F. Larry, Brenda L. Ekstrom, Janet Wanzek, and Timothy L. Mortensen. 1989. *Outlook of North Dakota Farm Households: Results of the 1988 Longitudinal Farm Survey*. Agricultural Economics Report No. 144. Fargo: North Dakota State University, Department of Agricultural Economics.

Mortensen, Timothy L., F. Larry Leistritz, Jay A. Leitch, and Brenda L. Ekstrom. 1988. *A Baseline Analysis of Participants in the Conservation Reserve Program in North Dakota*. Agricultural Economics Miscellaneous Report No. Fargo: North Dakota State University, Department of Agricultural Economics.

Mortensen, Timothy L., Jay A. Leitch, F. Larry Leistritz, Brenda L. Ekstrom, and Randal C. Coon. 1989. "An Analysis of Baseline Characteristics and Economic Impacts of the Conservation Reserve Program in North Dakota." Paper for presentation at Conference on the Social, Economic, and Environmental Consequences of the Conservation Components of the Food Security Act of 1985. Columbus, Ohio, March 1-2, 1989.

Taylor, Fred R., Laurel D. Loftsgard, and LeRoy W. Schaffner. 1961. *Effects of the Soil Bank Program on a North Dakota Community*. Agricultural

Economics Report No. 19. Fargo: North Dakota State University, Department of Agricultural Economics.
U.S. Department of Agriculture. March 1988. Conservation Reserve Program Statistics. Washington, D.C.: U.S. Government Printing Office.

Table 7.1 Business Sectors Affected by the Conservation Reserve Program and Items Purchased in Each Sector

Sector	Items Purchased
(8) Retail	Fertilizer, fuel, oil, seed, chemicals, machinery, hardware.
(9) Finance, insurance, and real estate	Crop insurance, property insurance, interest on borrowed capital.
(10) Business and personal services	Machinery repairs, custom farm operations, legal and accounting services.
(12) Households	Net income from farm operations, payments to hired labor.

Table 7.2 Selected Characteristics of CRP Land and Participants, North Dakota, 1988.

Item	Units	Value
Yields--CRP land compared to land not in CRP	Percent	-9.5
Input costs--CRP land compared to land not in CRP	Percent	0.5

(continued)

Table 7.2 Selected Characteristics of CRP Land and Participants, North Dakota, 1988.

Item	Units	Value
CRP contract payment compared to cash rent	Percent	6.7
Costs per acre to establish CRP cover	Dollars	37.20
Costs per acre to maintain CRP cover	Dollars	6.92
Annual CRP contract payment	Dollars	36.98
Type of CRP cover:		
Grass and/or legumes	Percent	91.0
Trees (on part of area)	Percent	9.0
Landowner Age	Years	57.2
Landowner residence:		
North Dakota	Percent	90.0
Bordering states	Percent	4.2
Elsewhere	Percent	5.8
Landowner occupation:		
Farmer	Percent	73.0
Other	Percent	27.0
Gross farm income, 1987 (farmers only):		
Average	Dollars	92,440
Distribution:		
Less than $40,000	Percent	34.5
$40,000 to $99,999	Percent	35.0
$100,000 to $249,999	Percent	23.3
Over $250,000	Percent	7.1
Net Cash Farm Income, 1987 (farmers only):		
Average	Dollars	
Distribution:		
Negative	Percent	14.2
$0 to $9,999	Percent	37.5
$10,000 to $19,999	Percent	17.2
$20,000 to $39,999	Percent	19.9
$40,000 and over	Percent	11.1

(continued)

Table 7.2 Selected Characteristics of CRP Land and Participants, North Dakota, 1988.

Item	Units	Value
CRP payment as a percent of net farm income:		
Over 100 percent or net farm income was negative	Percent	40.6
50 to 100 percent	Percent	13.2
26 to 50 percent	Percent	18.5
0 to 25 percent	Percent	27.8
Did the CRP program enable you to continue farming?		
Yes	Percent	20.6

Table 7.3 Acres Enrolled in CRP and the Associated Loss of Production Expenditures and Change in Income, by CRP Pool Group, 1987

Pool Group	Acres Through 5th Sign-up	Reduced Expenditures			Change in Income
		Retail(8)	FIRE(9)*	B&P Serv(10)	Households(12)
	(thousand dollars)				
1	244,518	-4,940	-1,787	-1,619	10
2	381,409	-8,539	-3,074	-2,649	2,033
3	260,548	-6,563	-2,406	-1,961	755
4	240,997	-7,986	-2,541	-1,950	-92
5	174,975	-7,262	-2,112	-1,772	-1,448
State	1,302,048	-35,291	-11,919	-9,951	1,258
State Total Percentage of Reduced Expenditures		61.7%	20.9%	17.4%	

* Finance, Insurance and Real Estate

Table 7.4 Average 1980-1987 Baseline Business Activity and Business Activity Associated With Reduced Production Expenditures and Income Change Resulting From CRP Acres by Economic Sector, North Dakota, 1987

Sector	Baseline Business Activity[a]	CRP Business Activity		
		Production Expenditures	Income Change	Net Change
	----------------thousand dollars----------------			
(1) Ag, livestock	1,406,058	-4,254	85	-4,169
(2) Ag, crops	3,662,184	-1,709	33	-1,676
(3) Nonmetal mining	49,420	-186	7	-179
(4) Construction	730,076	-2,650	113	-2,537
(5) Transportation	91,330	-627	12	-615
(6) Comm & pub utilities	659,314	-4,540	133	-4,407
(7) Ag proc & misc mfg	2,143,329	-2,670	52	-2,618
(8) Retail trade	5,321,801	-57,505	937	-56,568
(9) FIRE*	1,110,927	-16,731	211	-16,520
(10) Bus & pers services	488,715	-12,056	76	-11,980
(11) Prof & soc services	521,151	-2,442	124	-2,318
(12) Households	7,955,811	-35,685	1,953	-33,732
(13) Government	679,028	-3,437	136	-3,301
(14) Coal mining	134,774	0	0	0
(15) Thermal elec generation	225,900	0	0	0
(16) Petroleum exp/extraction	883,623	0	0	0
(17) Petroleum refining	120,864	0	0	0
TOTAL	26,247,305	-144,492	3,872	-140,620

[a] Baseline business activity is based on the 1980-1987 average sales for final demand in terms of 1987=base dollars.

* Finance, Insurance and Real Estate

Table 7.5 Distribution of CRP Acres, Total CRP Impact on Business Volume, and CRP Related Employment Change Among Pool Groups

Pool Group	CRP Acres	Total CRP Impact	CRP Impact as a Percentage of Pool Baseline	CRP-Related Employment Change
	(%)	(million $)	(%)	(number)
1	18.8	21.2	-0.33	371
2	29.3	30.0	-0.68	552
3	20.0	25.5	-0.52	453
4	18.5	31.6	-0.91	523
5	13.4	32.2	-0.39	517
TOTAL	100.0	140.5	-0.54	2,416

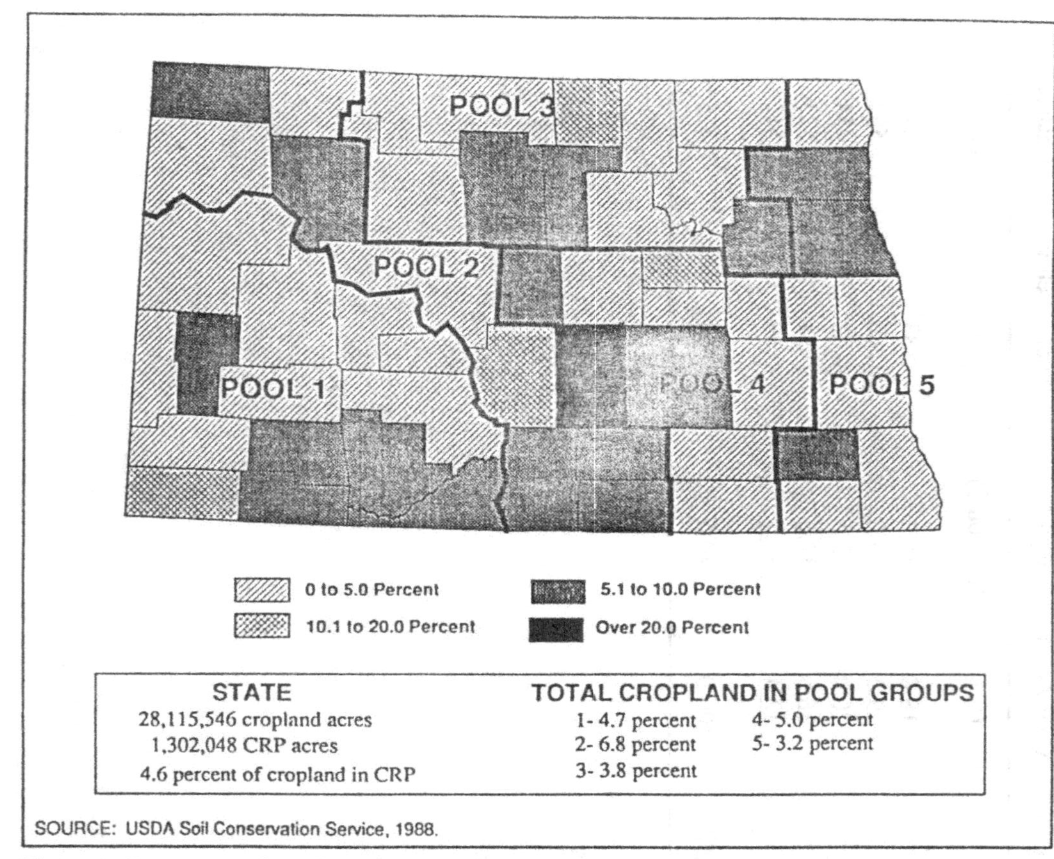

Figure 1. Percentage of total cropland enrolled in CRP by category, July 1987.

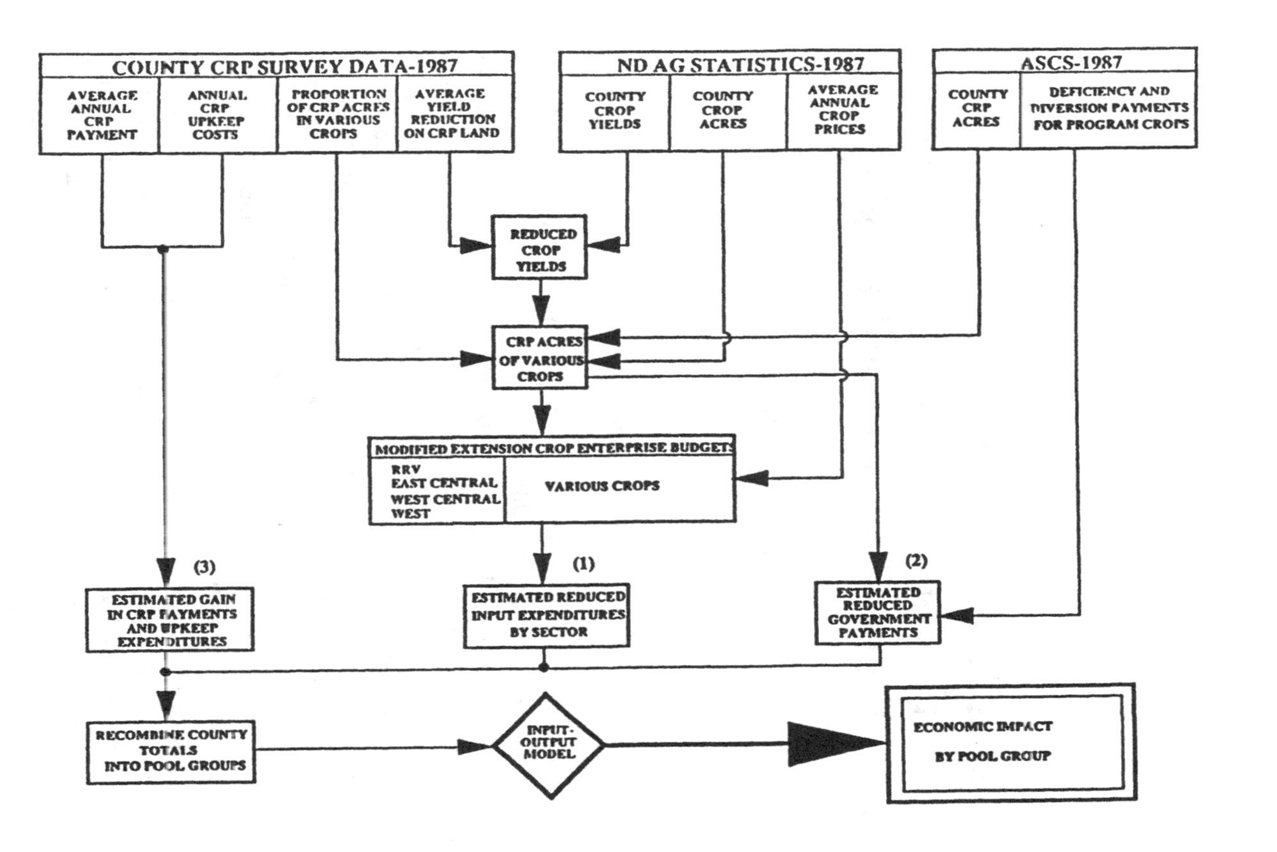

Figure 2. Procedure for estimating direct impacts of CRP.

8

Using Input-Output Analysis for Estimation of Distributional Impacts from Plant Openings and Closings

Thomas R. Harris

Introduction

With current changes occurring in rural America, local economic development groups have expressed interest in estimating the economic impacts of proposed recruitment of industrial plants or closing of an industrial plant in their community. With recruitment of industrial plants, community economic development groups are aware of the costs of recruitment, i.e. building of industrial parks, vocational education programs, tax incentives, etc., however, local economic development groups may be unable to estimate the potential impacts of the incoming plant. Conversely, local economic development groups aware of possible closing of a local industry may want information as to the local economic impacts of the existing firm in order to develop mitigation plans from the proposed industrial plant closing.

One method of determining the impacts of industrial plant openings and/or closings is input-output analysis. Several procedures have been developed to analyze these impacts using input-output analysis such as a final demand approach, expansion of existing sector capacity, and the location of an industrial sector currently not present in the local economy. Isard and Kuenne (1953) and Miller (1957) developed procedures to estimate impacts of in-movements of industrial plants into a region. Adams (1974) completed a study in the Lower Rio Grande Valley of Texas in which an interindustry model was used to determine the impacts to the regional economy from a new or expanding industry in the study area. Guedry and Smith (1980) later developed a methodology to estimate the distributional impacts in a rural Louisiana economy from in-movement of a manufacturing plant. Miller and Blair (1985) present a number of procedures and examples to analyze industrial plant openings and/or closings in a local economy.

Therefore, the primary objective of this paper is to show how input-output models can be used to estimate impacts to a regional economy from industrial plant openings and/or closings. The paper will be divided into three sections to meet this objective. First, the procedures developed by Guedry and Smith (1980) to estimate the distributional impacts of industrial plant location will be presented. Second, an example industrial plant location in the northeastern Nevada counties of Humboldt and Lander will be developed. Finally, the use of input-output modeling procedures for impact analysis from industrial plant closings will be discussed.

Mathematical Description

Interindustry analysis was developed by Wassily Leontief in 1936. In general, the transactions of a regional economy can be states as:

$$(1) \quad X = AX + Y$$

where X is an (nx1) matrix that includes the total output for n sectors, A is an (nxn) matrix of technical coefficients that are determined by dividing purchases of a given sector from other sectors in the regional economy by the given sector's total output, and Y is an (nx1) matrix of final demand for each of the n sectors. The matrix can be rewritten as:

$$(2) \quad X(I-A) = Y$$

where I is the identity matrix and when the (I-A) matrix or Leontief matrix is inverted, the inversion results in the interindustry matrix:

$$(3) \quad X = (I-A)^{-1} Y$$

The interindustry matrix can be used to determine the regional economic impacts from adjustments in a given sector. Adjustments are increased sales to final demand by a given sector or decreases in the level of sectoral output. However, to derive the distributional economic impacts of a new industry to a local economy procedures developed by Guedry and Smith (1980) are used.

To derive the impacts of a new industry or sector to a local economy, the technical coefficients matrix is portioned as:

$$(4) \quad A = \frac{A_{11} \quad A_{12}}{A_{21} \quad A_{22}}$$

where there are p sectors whose distributional impacts are to be measured or:

A_{11} is an (n-p) X (n-p) matrix of technical coefficients,

A_{12} is an (n-p) X p matrix of technical coefficients,

A_{21} is a p x (n-p) matrix of technical coefficients, and

A_{22} is a p x p matrix of technical coefficients.

Given the portioned matrix in equation 4, equation 3 can be rewritten as:

$$(5) \quad \frac{X_1}{X_2} = \frac{(I-A_{11}) \quad -A_{12}}{-A_{21} \quad (I-A_{22})}^{-1} \frac{Y_1}{Y_2}$$

The impacts of a new sector are given by X_2 and Y_2. If the model only estimates the distributional impacts on other sectors in the regional economy from the location or relocation of a new industry, equation 5 becomes:

$$(6) \quad \frac{X_1}{X_2} = \frac{(I-A_{11}) \quad 0}{0 \quad I} \frac{Y_1^*}{Y_2^*}$$

The final demand column denoted as Y^* represents final demands of the p sectors which are removed from the endogenous portion of the model. The inverse in equation 6 is obtained postmultiplying the A matrix by an (nxn) matrix of ones except for the pth elements. The inverse of equation 5 becomes:

$$(7) \quad (I-J_pAJ_p)^{-1}$$

where:

$$(8) \quad J_p = \frac{I_{(n-p) X (n-p)} \quad 0_{(n-p)xp}}{0_{px(n-p)} \quad 0_{pxp}}$$

so:

$$(9)\quad J_pAJ_p = \begin{array}{cc} A_{11} & 0 \\ \hline 0 & 0 \end{array}$$

From equations 4 and 5 the distributional impacts of the new sector p is given by:

$$(10)\quad D = (I\text{-}A)^{-1} - (I\text{-}J_pAJ_p)^{-1}$$

Where D is an (nxn) matrix of distributional coefficients which identify the part of indirect output generated in sector i by a change in final demands of sector j as a result of the interrelationships generated by the new industry.

Example Analysis of the Distributional Impacts of a Plant Opening in a Local Economy

For this paper, the input-output model for the northeastern counties of Humboldt and Lander will be used to estimate local economic impacts of a new industry.[1] The counties of Humboldt and Lander have gone through significant change in their production of agricultural crops. For many years, the value of agricultural output for these two counties was relatively small when compared to values for other Nevada counties. However, with the introduction of potato farming in the late 1960's, the value of agricultural production for these counties has increased significantly. With current potato production in Humboldt and Lander counties and no potato processing plant in the state, the local economic development authority targeted the location of a potato processing plant in Humboldt and Lander counties as a primary objective. With the eminent construction and operation of the potato processing plant, the local economic development group wanted information as to the potential economic impacts of this new industry. The procedures as outlined by Guedry and Smith (1980) were used to estimate the distributional impacts of the location of the potato processing plant in the two-county economy.

For this analysis, the Household Sector was included in the processing section of the transactions table so the direct, indirect, and induced impacts of sectoral sales to final demand are derived. Table 8.1 shows the direct, indirect, and induced requirements of sectors in the Humboldt and Lander economy with the inclusion of the Potato Processing Sector. From Table 8.1, the impacts to the Humboldt and Lander economy from a one dollar sale to final demand by the Potato Processing Sector is $2.13 or when the Potato Processing Sector increases sales to final demand by $1.00, the economic activity in the economies of Humboldt and Lander counties would increase by $2.13. By scanning the Potato Processing Sector row (row 11), the interdependencies of the Potato

Processing Sector with other sectors in the regional economy can be estimated. Interdependency coefficients in row 11 show the direct, indirect, and induced output requirements created in the Potato Processing Sector from changes in final demand sales by other sectors in the two-county economy. Also, from Table 8.1, the interdependencies of the Potato Processing Sector are the largest of all sectors in the two-county economy. This is a usual result for the agricultural processing sectors because they purchase a large proportion of their input locally and sell a large proportion of their output to final demand (export sales).

Distributional Impacts of the Potato Processing Sector

The matrix of the distributional impacts of the incoming industry derived from equation 10 are shown in Table 8.2. Coefficients from the $(I-A)^{-1}$ matrix (Table 8.1) represent the direct, indirect, and induced impacts of the two-county economy with the Potato Processing Sector included, while coefficients of the $(I-J_pAJ_p)^{-1}$ matrix represent the direct, indirect, and induced impacts of the two-county economy exclusive of the Potato Processing Sector. Differences in the $(I-J_pAJ_p)^{-1}$ matrix and the $(I-A)^{-1}$ matrix estimate the distributional impacts on the two-county economy from the location of the Potato Processing Sector in Humboldt and Lander Counties.

Coefficients in Table 8.2, with the exception of coefficients in the Potato Processing Sector (column 11), estimate the increase in each sector's indirect and induced output that can be attributed to the location of the Potato Processing Sector by other sectors in the regional economy. For example, $0.00001 of additional output is gained by the Transportation, Communication, and Utilities Sector from a $1.00 increase in sales by the Agricultural Sector as a result of increased purchases by the Potato Processing Sector. The column summation of the Agricultural Sector (sector 1) shows that from a $1.00 increase in sales to final demand by the Agricultural Sector, an additional $0.00019 in indirect and induced output demand is created in the two-county economy resulting from the trade generated by the purchases of the newly located Potato Processing Sector. The largest distributional impact that the Potato Processing Sector has on sectors in the two-county economy is the Transportation, Communications, and Utilities Sector. From trade created by purchases of the Potato Processing Sector, $0.00130 of additional output in the economy from a $1.00 increase in sales to a final demand by the Transportation, Communications, and Utilities Sector is realized from the location of the Potato Processing Sector.

By looking down the Potato Processing Sector column (column 11) in Table 8.2, the impacts on other sectors of the two-county economy from sales to final demand by the Potato Processing Sector are presented. As for the Agricultural Sector, $0.57823 of increased output is required from the Agricultural Sector in response to a $1.00 increase in sales to final demand by the incoming Potato Processing Sector. The summation of column 10 in Table 8.2 estimates that $1.13 of regional output is created indirectly and induced because of the location of the Potato Processing Sector.

Plant Closing Analysis

Deriving impacts of the out-movement of a firm on the entire economic sectors of a local economy can be treated in some degree with the same procedures as the analysis of an incoming industry or sector. Usually final demand, output, income, and employment multipliers provide an adequate approach to quantifying such decreases in economic activity - particularly if one plant closes, but other plants in the same sector remain in production. If a complete sector leaves a local economy, procedures as outlined by Guedry and Smith (1980) could be used to derive the distributional impacts in the local economy of the leaving sector. However, additional analysis for a leaving sector may be required such as the determination that local sectors will either import inputs that have disappeared from the local economy or else substitute alternative locally produced inputs. Similarly, local firms that previously supplied inputs to the now absent sector will find their sales patterns altered. Again, changes will occur in the technical coefficients matrix and the precise estimation of these changes and the magnitudes of these changes will be difficult to estimate.

Conclusion

Input-output model provides a framework within which to assess the economic impacts associated with a new industry locating in a local economy or an industry leaving a community. Procedures outlined by Guedry and Smith (1980) were used to estimate distributional impacts of a new industry locating in the northeastern Nevada counties of Humboldt and Lander. Procedures used to estimate distributional impacts of a new industry could also be applied to a firm leaving a local economy.

Procedures used in this paper stressed estimation of the distributional impacts of a locating firm. However this analysis can be extended to derive levels of regional economic expansion from location of a firm through the final demand approach or assuming the new firm's level of output is exogenously derived. With sectoral output of the new industry exogenously derived, sectoral production levels for all sectors in the regional economy can be derived as well as incorporation of sectoral production capacities.

Notes

[1] The input-output model for Humboldt and Lander counties used in this paper was developed by Fillo et al. (1978).

References

Adams, John W. *Economic Impact of New or Expanded Industries in the Lower Rio Grande Region of Texas.* Office of Information Services, Austin, Texas, MP-1129, April 1974.

Fillo, Frank O., Hans O. Radtke, and Eugene P. Lewis. The Economy of Humboldt and Lander Counties: A Working Model for Evaluating Economic Change. Mimeo Series MS-99, Department of Agricultural Economics, University of Nevada Reno, 1978.

Guedry, Leo T. and David W. Smith. "Impact of Industry in Rural Economies: An Input-Output Approach." *Southern Journal of Agricultural Economics.* 12(1980): 19-24.

Isard, Walter and Robert E. Kuenne. "The Impact of Steel Upon the Greater New York-Philadelphia Industrial Region." *Review of Economics and Statistics*, 35(1953): 289-301.

Leontief, Wassily W. "Quantitative Input-Output Relations in the Economic System of the United States." *Review of Economics and Statistics*, 18(1936): 105-125.

Miller, Ronald E. "The Impact of the Aluminum Industry on the Pacific Northwest: A Regional Input-Output Analysis." *Review of Economics and Statistics.* 39(1957): 200-229.

Miller, Ronald E. and Peter D. Blair. *Input-Output Analysis: Foundations and Extension.* Prentice-Hall, Inc., Englewood Cliffs, NJ, 1985.

Table 8.1. Direct, Indirect, and Induced Trade Requirements, Humboldt and Lander Counties, Nevada.

Sectors	1 Agricul- ture	2 Mining	3 Manufac- turing	4 Construc- tion	5 Transpor- tation	6 Trade
1	1.03175	0.00094	0.00061	0.00139	0.00115	0.00198
2	0.00000	1.00000	0.00000	0.00000	0.00000	0.00000
3	0.00264	0.00192	1.00198	0.04763	0.00216	0.01078
4	0.01717	0.01712	0.01195	1.12439	0.01584	0.03196
5	0.08294	0.08624	0.01723	0.04442	1.02268	0.05196
6	0.07697	0.02205	0.01871	0.04460	0.02302	1.04971
7	0.04120	0.01434	0.01279	0.04858	0.01501	0.03654
8	0.07480	0.03843	0.03296	0.09763	0.03850	0.07077
9	0.03142	0.02928	0.00694	0.02322	0.02596	0.02037
10	0.23271	0.24745	0.16051	0.35190	0.21470	0.46762
11	0.00009	0.00007	0.00002	0.00005	0.00061	0.00048
TOTAL	1.59170	1.45784	1.26370	1.78381	1.36507	1.74663

Sectors	7 FIRE*	8 Services	9 Local Gov't	10 House- hold	11 Potato Processing
1	0.00133	0.00287	0.00214	0.00423	0.57823
2	0.00000	0.00000	0.00000	0.00000	0.00001
3	0.00527	0.00843	0.00968	0.00794	0.00280
4	0.02419	0.02075	0.04134	0.07756	0.02143
5	0.03298	0.04483	0.05235	0.08021	0.10213
6	0.03149	0.03368	0.04861	0.09793	0.05410
7	1.02892	0.02901	0.03565	0.06518	0.03550
8	0.05903	1.05765	0.11642	0.15258	0.06373
9	0.02756	0.01545	1.01426	0.02812	0.02495
10	0.35761	0.28712	0.51181	1.15778	0.24689
11	0.00004	0.00004	0.00036	0.00010	1.00009
TOTAL	1.56842	1.49983	1.83259	1.67164	2.12985

Source: Input/Output Model for the State of Nevada, University of Nevada, Reno.
* Finance, Insurance and Real Estate

Table 8.2. Direct, Indirect, and Induced Trade Requirements, Humboldt and Lander Counties, Nevada.

Sectors	1 Agriculture	2 Mining	3 Manufacturing	4 Construction	5 Transportation	6 Trade
1	0.00003	0.00004	0.00001	0.00003	0.00035	0.00028
2	0.00000	0.00000	0.00000	0.00000	0.00000	0.00000
3	0.00000	0.00000	1.00000	0.00000	0.00000	0.00000
4	0.00000	0.00000	0.00000	0.00000	0.00001	0.00001
5	0.00001	0.00001	0.00000	0.00001	0.00001	0.00005
6	0.00000	0.00000	0.00000	0.00000	0.00003	0.00003
7	0.00000	0.00000	0.00000	0.00000	0.00002	0.00002
8	0.00001	0.00000	0.00000	0.00000	0.00004	0.00003
9	0.00000	0.00000	0.00000	0.00000	0.00002	0.00001
10	0.00002	0.00002	0.00001	0.00001	0.00015	0.00012
11	0.00009	0.00007	0.00002	0.00005	0.00061	0.00048
TOTAL	0.00019	0.00014	0.00004	0.00010	0.00124	0.00103

Sectors	7 FIRE*	8 Services	9 Local Gov't	10 Household	11 Potato Processing
1	0.00002	0.00003	0.00021	0.00006	0.57823
2	0.00000	0.00000	0.00000	0.00000	0.00001
3	0.00000	0.00800	0.00000	0.00000	0.00280
4	0.00000	0.00000	0.00001	0.00000	0.02143
5	0.00000	0.00000	0.00004	0.00001	0.10213
6	0.00000	0.00000	0.00002	0.00001	0.05410
7	0.00000	0.00000	0.00001	0.00000	0.03550
8	0.00000	0.00000	0.00002	0.00001	0.06373
9	0.00000	0.00000	0.00001	0.00000	0.02495
10	0.00001	0.00001	0.00008	0.00002	0.24689
11	0.00004	0.00004	0.00036	0.00010	0.00009
TOTAL	0.00007	0.00008	0.00076	0.00021	1.12985

Source: Input/Output Model for the State of Nevada, University of Nevada, Reno.
* Finance, Insurance and Real Estate

9

Impacts of Transfer Payments

G. Andrew Bernat, Jr.

Introduction

Transfer payments make up a growing share of total personal income in the United States. In 1960, transfer income comprised 7 percent of personal income. By 1970, the share had increased to 10 percent, and during the 1980's, over 14 percent of personal income consisted of transfer payments (Council of Economic Advisors). As transfer payments increase over time -- at the regional level as well as at the national level -- it becomes increasingly important to estimate the impacts of changes in the level of transfer payments on regional economies.

The purpose of this paper is to demonstrate how the impacts of changes in transfer payments can be estimated using input-output models. In the first part of the paper some of the problems encountered in estimating the impact of changes in transfer payments are discussed. The regions and the models used in the study are then described, followed by a discussion of the results.

Transfer Payments in an I/O Framework

Household income in a regional I/O model is divided into two components: endogenous income and exogenous income. Endogenous income includes wages, salaries, and proprietors' income -- essentially any income arising directly from the economic activity within the region. Exogenous income comprises all income not considered endogenous. Examples of exogenous income in a regional framework are transfer income, interest and rental payments on assets held outside the region, and dividend payments from extra-regional firms and businesses. Transfer payments consist primarily of Social Security payments, unemployment payments, Aid to Families with Dependent Children, and other public assistance payments. Two aspects of the transfer portion of exogenous income are of interest in this study: the distribution of transfers, and their endogenous component.

The Distribution of Transfers

One of the most obvious characteristics of income transfers is their distribution across households. Rather than being distributed uniformly they are generally concentrated among lower income households. If all households exhibited similar expenditure patterns, this concentration would be irrelevant. However, because household expenditures vary with income, the only way to properly model the impacts of a change in transfers is to explicitly account for differences in expenditure patterns.

In many regional input-output models, households are modeled as a single sector. Estimated impacts of a change in exogenous income will depend only on the size of the change. Hence, such a model would indicate that a $1 million change in Social Security payments would have exactly the same impact as a $1 million change in interest and dividend payments. However, as the results of this study indicate, the distribution of payments can make a substantial difference in the regional impacts.

Endogenous Transfers

The second characteristic of income transfers that needs to be addressed is the fact that some income transfers are not entirely exogenous in that the levels of payment are tied either to employment status or income. Because most of these transfers are inversely related to income, ignoring their connection to regional employment results in estimates of impact multipliers that are too high.

As an example, consider what happens if the output of a sector increases, thereby increasing regional employment and household income. Assume that some of the new jobs are filled from the pool of unemployed workers in the region. Presumably these workers had some source of income while unemployed that enabled them to maintain at least a minimal level of expenditures on regionally produced goods and services. If this income had been in the form of a transfer payment tied to income, as many transfers are, the transfer income would cease, or at least be reduced, when the worker gained employment. The impact on the regional economy would be the difference between the pre-employment income and the post-employment income. If, on the other hand, the transfer payment is ignored, the estimated impact would consist of the entire wage and salary payments generated by the increase in sectoral output.

Unless we are willing to assume either that all new jobs are taken by in-migrants or that unemployed workers have zero expenditures, the level of transfer payments that are means tested or related to employment status must be adjusted for every change in regional output. In contrast to standard regional input-output models, the models used in this study incorporate an attempt at endogenizing some of this type of transfer payment.

Modeling Transfers

As shown above, an input-output model must have two characteristics to properly handle income transfers. First, the household sector needs to be disaggregated according to household income in order to properly capture the effects of the concentrated nature of most income transfers. Second, because certain transfers are not truly exogenous, payment levels should be tied to the level of regional economic activity.

Conceptually, disaggregation of the household sector is straightforward. Following Miyazawa (1976)[1], an input-output model which explicitly accounts for different income classes can be represented by the following system of equations:

$$(1) \quad \begin{bmatrix} x \\ y \end{bmatrix} = \begin{bmatrix} A & C \\ V & 0 \end{bmatrix} \begin{bmatrix} y \\ x \end{bmatrix} + \begin{bmatrix} f \\ g \end{bmatrix}$$

The n by n matrix A and the vectors x and f represent the interindustry coefficient matrix, the output of n regional producing sectors, and exogenous demand for regional output, respectively. The n by k matrix C indicates how each of the k household groups' expenditures are distributed over the n regional producing sectors. The k element vector g represents exogenous income accruing to regional households. The k by n matrix V distributes the value added generated by each of the n regional producing sectors to the k household groups, where households have been grouped according to household income.

The V matrix is in turn defined by:

$$(2) \quad V = QWFx$$

Each element of the s by n matrix F indicates the number of employees in each of the s occupational groups per dollar of output of each of the n regional producing sectors. The product Fx, an s by n matrix, therefore indicates the total number of workers, by occupation, required by the output vector x. W is an s by s matrix of value-added coefficients indicating the value-added associated with each occupation. The k by s matrix Q is the household-occupation matrix. Each column of Q allocates the workers of a given occupation to households in each of the k household groups. In essence, each element of Q is the probability that a worker of a particular occupation will be in a household of a particular income class.

Solving for regional income and regional output levels results in:

$$(3) \quad \begin{bmatrix} x \\ y \end{bmatrix} = \begin{bmatrix} (I - A) & -C \\ -V & I \end{bmatrix}^{-1} \begin{bmatrix} f \\ g \end{bmatrix}$$

The next step, that of endogenizing transfer payments that are tied to income or employment status is more difficult than disaggregating the household sector. The approach taken here is to assume that every person in the work force who is unemployed receives either unemployment benefits or other income transfers that can be approximated by unemployment benefit rates. These transfers cease when the unemployed worker becomes employed.

If the vector z represents the number of unemployed workers, by occupation, then total unemployment benefit payments is equal to Uz, where U is a diagonal matrix of unemployment benefit payment rates. Since the number of unemployed workers is simply the difference between the total labor force and the number of employed workers, unemployment benefit payments can be expressed as:

$$(4) \quad Uz = U(s - Fx)$$

where s is a vector of labor supply by occupation and F and x are defined as above.

Redefining g in the income block of equation (1) so that it does not include unemployment payments and using (4) to represent the level of unemployment benefits, household income by income class becomes:

$$(5) \quad y = Vx + U(s - Fx) + g$$

Substituting the definition of V given in (3) and rearranging results in:

$$(6) \quad y = Q(W-U)Fx + (Us + g)$$

The matrix V in (3) is therefore redefined to be equal to Q(W-U). As (Us + g) is unrelated to the level of regional economic activity, we have accomplished our goal of endogenizing the transfer payments that are tied to regional employment.

There are a number of shortcomings in the above procedure. Chief among them is the narrow definition of the unemployment benefit payment rate matrix, U. In the above models, the elements of U are based on the unemployment benefit schedule. This implies that all workers receive unemployment benefits while unemployed. A more correct construction of U would be based on all transfer programs that would be available to unemployed workers. However, in the absence of this more correct implementation of U, unemployment program payments will be used as a proxy for the other means tested income transfers.

The models, one for each of the three defined regions in Virginia, were constructed entirely from secondary sources. The technical coefficients matrix was constructed from a 72 sector U.S. model. Regional purchase coefficients, developed by the Regional Science Research Institute (RSRI) were used to convert the national coefficients to regional coefficients. The household expenditure coefficients were based on the Consumer Expenditure Survey of BLS in conjunction with the household expenditure column of an I/O model of Virginia developed by RSRI. The 90 by 72 matrix of employment coefficients, F, was constructed by RSRI. The matrices W and U are diagonal matrices. W has wage, salary, and proprietors income rates while the diagonal elements of U are unemployment benefit rates. Both matrices were constructed using the Census' Public Use Microdata Sample (PUMS).[2]

The Study Regions

Three regions of Virginia were chosen in order to examine diverse economies. Two of the regions are primarily rural: one is primarily a farming region while the other is dominated by the coal mining industry. The third region is largely urban. The populations, according to the 1980 Census of Population (Bureau of the Census), were fairly close. The urban region had 85,600 households, the coal region 78,100 households, and the farming region had 56,700 households.

Average household income in the urban region was $21,024. This was nearly 50 percent higher than in the other two regions. The distribution of income was relatively more skewed in the rural regions than the urban region. In the coal region, 31 percent of all households were in the lowest income group and accounting for less than 6 percent of regional income while the 1.4 percent of all households that were in the highest income class received over 7 percent of all income. In the urban region, 21 percent of all households were in the lowest income group, accounting for 1.6 percent of all income and the 3.3 percent of households in the highest income class received over 11 percent of all income.

In all three regions public assistance payments and social security payments were concentrated in the lower income groups but comprised a much smaller proportion of total income in the urban region than in the other regions. In the coal region, public assistance payments accounted for 8 percent of total income and social security 16 percent. Each transfer category comprised roughly 6 percent of household income in the farming region. In contrast, only 3 percent of the urban region's household income consisted of public assistance payments and 2.4 percent social security.[3]

While both types of transfer income make up a larger portion of total income in the coal region than the farming or urban regions, public assistance payments are more concentrated in the latter two regions. Over half of all public assistance payments went to households in the lowest income class in both of these regions compared to less than 40 percent in the coal region.

Results

In order to see how changes in transfers affect the three regions, Social Security payments and public assistance payments were each increased by ten percent. The results are summarized in Table 9.1.

Two factors determine the size of the impacts: the level of transfers and the strength of linkages. In the case of the regions examined here, the dominant effect is the size of transfer income. The largest impacts occurred in the coal region for both types of transfer. The total impacts on household income in the coal region of a 10 percent increase in Social Security income would be almost 80 percent larger than the impact in the farming region and over 40 percent larger than the impact in the urban region. The output impacts were also larger in the coal region than the other regions but by a smaller margin. However, most of the difference in the size of impacts is caused by the difference in the size of the initial change in transfers.

In order to better see the importance of the distribution of transfers, the models were used to estimate the impacts of a $1 million increase in both types of transfers. In addition, an increase of the same size was distributed according to the distribution of interest and dividend income. The results are summarized in Table 9.2.

In all three regions, total impacts are the largest for Social Security payments and lowest for dividend and interest payments. The impacts of public assistance payments were lower than for Social Security but were similar in their distribution across sectors. In the coal region, the change in total output due to a change in interest and dividend payments would be 80 percent of an equal change in public assistance payments and only 75 percent of the output change due to an equal increase in Social Security payments. Likewise, income generated by a $1 million change in interest and dividend payments would be only 77 percent of the income generated by an equal change in Social Security payments and 82 percent of the income generated by a change in public assistance payments.

The relatively low showing for interest and dividend payments can be attributed to the fact that this income component is concentrated among the higher income households. The primary reason higher income households have lower multipliers is that these households spend a smaller proportion of each dollar received than do lower income households.

The second reason is the difference in the pattern of linkages between upper income households and other households. If the spending patterns of high and low income households differed only in the total marginal propensity to consume, then the sectoral impacts of a change in interest and dividend income would be proportional to the impacts of a change in Social Security income. However, because the pattern of expenditure differs in addition to differences in marginal propensities to consume, sectoral impacts are not proportional. In the coal region, for example, even though a change in Social Security payments results in a total impact on output nearly 30 percent larger than an equal change in interest and dividend payments, the output of five sectors would be higher after a change in interest and dividend payments than after a change in transfer payments. The distribution of value-added for two of these five sectors--educational services and hotels and lodging -- were more skewed towards high income households than the average. The more pervasive is this type of linkage--where high income households are strongly linked to sectors which are in turn linked to high income households -- the greater will be the differences in the distribution of impacts across sectors.

Summary and Conclusions

Two major conclusions can be drawn from the results. First, it is important to use a model with a disaggregated household sector when modeling the impacts of changes in transfer income. The results shown here indicate that, depending on the region, substantial differences in multipliers exist between the two types of models. However, the greater the degree of linkage within an economy, the smaller the differences are likely to be.

The second conclusion is that some mechanism needs to be developed that will endogenize the portions of transfer income which are tied to regional activity. While the method employed in this study may have its shortcomings, it nevertheless is a first step in an important direction. If such an adjustment is not made, impacts will be overestimated unless we are willing to assume that all newly employed workers are in-migrants and all newly unemployed workers leave the region.

Notes

[1] For another recent application of Miyazawa's framework see Rose and Beaumont (1988).

[2] For more detail on the construction of these models see Bernat.

[3] These estimated shares are higher than the actual shares because they were taken from PUMS. The maximum income level in the PUMS data is $75,000. Consequently, the share of the highest income class is underestimated and the shares of all lower income classes is overestimated. However, the shares should give a relatively accurate view of the differences across regions.

References

Bernat, G. Andrew Jr. "Income Distribution in Virginia: The Effect of Intersectoral Linkages on the Short-Run Size Distribution of Income in Small Regions," Virginia Polytechnic Institute and State University, July 1985.

Bureau of the Census. *Public Use Microdata Samples Technical Documentation.* U.S. Department of Commerce, March 1983.

Council of Economic Advisors. *Economic Report of the President.* U.S. Government Printing Office, January 1989.

Miyazawa, K. *Input-Output Analysis and the Structure of Income Distribution.* New York: Springer-Verlag, 1976.

Rose, Adam and Paul Beaumont, "Interrelational Income-Distribution Multipliers for the West Virginia Economy." *Journal of Regional Science,* 28 (1988): 461-475.

Table 9.1. Impacts of a 10 Percent Change in Transfer Income (in $ million)

Region	Impacts		
	Initial Change	Income	Output
Coal Region			
Social Security	9.1	10.4	4.2
Public Assistant	17.7	20.2	7.2
Farm Region			
Social Security	5.2	5.8	3.3
Public Assistant	5.2	5.7	2.5
Urban Region			
Social Security	5.7	6.8	3.6
Public Assistant	4.4	5.1	2.3

Source: Input/Output Model for Virginia, Virginia Tech.

Table 9.2. Comparison of the Impacts of a $1 Million Change in Transfer Income (in percentages)

Region	Ratio of Impacts to SS Impacts	
	Income	Output
Coal Region		
Interest and Div.	77	75
Public Assistant	82	80
Farm Region		
Interest and Div.	80	83
Public Assistant	78	82
Urban Region		
Interest and Div.	84	84
Public Assistant	83	82

Source: Input/Output Model for Virginia, Virginia Tech.

10

Structural Analysis Using Input/Output Analysis: The Agriculture Sector and National and Regional Levels

Gerald Schluter

Introduction

The analyst using I/O for structural analysis at the national level has advantages over I/O analysts at the subnational level. At the national level, analysts often find data more clearly defined and often more available. This clarity and availability comes at a cost of model flexibility for the researcher. In this paper I will expand somewhat on this central theme and in the process mine some of our experiences doing I/O analysis at the national level for insights which may help I/O analysts at the sub-national level in their I/O-based impact analysis.

In the Economic Research Service, input-output analysis is used in three major roles - measuring the impact of agricultural trade, defining the U.S. Food and Fiber System (FFS), and measuring the economywide effects of agricultural policy proposals. Stated alternatively, measuring the impacts of a particular sector product, defining the size of that part of the economy that is linked forward and backward to a particular sector, and measuring the economywide effects of sector specific policies.

Trade Impact Analysis

Our trade impact analysis follows traditional I/O analysis procedures. Agricultural exports and imports are clearly final demands to the sector and the economy. We start with a particular year's actual agricultural trade. We summarize the commodity trade data into I/O sector groupings, convert these sector values from port value to producer value, allocate the margins to the appropriate sectors, deflate sector trade values with sectoral specific deflators, and use the 1977 national BEA I/O model to estimate output, income, and employment related to a particular year's trade. We then reinflate income and output estimates to current prices. In 1988 the U.S. exported $37.1 billion of agricultural commodities and imported $14.7 billion of competitive agricultural products. Associated with this level of exports was an additional $52.2 billion of supporting goods and services ($46.3 billion nonfarm) and 1,025,000 jobs (601,000 nonfarm jobs). An open model is used to estimate income from exports? A dollar of exports is expected to translate into a dollar of value-added or Gross National Product (gnp) - an income multiplier of one. In fact, our calculations imply an estimated $37.1 billion of exports generate $44.9 billion of gnp because our underlying assumption about prices wasn't true. Although our I/O impact analysis assumed no change in relative prices from the base year, 1977, Figure 1 illustrates, why this can be a shaky assumption for analysis of the agricultural sector.

Figure 1 presents agricultural relative prices (terms of trade) from two perspectives - value-added prices and output prices. The value-added terms of trade, defined as the ratio of the agricultural GNP deflator to the non-agricultural GNP deflator, is a measure of factor income per unit of real value added in agriculture relative to that in the non-agricultural sectors. A unit of real value added is a composite indicator of primary factor inputs --capital, labor, and land. If all factors are mobile and factor markets function perfectly, then each primary factor would earn the same return in all sectors. On average, the value-added terms of trade would then be constant over time. We do not observe such constancy, however. For agriculture, land is a sector-specific primary factor. Therefore, one would expect to find some variation in the value-added terms of trade as land values change over time. Except for land, changes in the value-added terms of trade indicate continuing disequilibrium in the factor markets and indicate "resource pull" effects that should lead to further factor reallocation. Figure 1 shows this series' movements over time.

The output terms of trade is the ratio of the agricultural prices received index to the non-agricultural GNP deflator. It indicates the purchasing power of agricultural goods in terms of non-agricultural domestically produced goods.

The output terms of trade have followed a trend similar to that of the value-added terms of trade (Figure 1), although there are some notable differences. During the Great Depression, the output terms of trade indices dropped at the same rate as the value-added terms of trade. However, the output terms of trade did not surpass its pre-depression high during the recovery of the mid-1930's. The output and value-added terms of trade rose together during the World War II boom, and fell only slightly before climbing again during the Korean War.

After the post-Korean War readjustment, the indices ceased to move together. The output terms of trade indices continued to decline, but the value-added indices leveled off. This divergence likely reflects a structural and productivity shift in agriculture.

Changing relative prices have implications for input-output analysis because of changed multipliers. In real (1977) dollars a million dollars of agricultural exports in 1988 generated $2.25 million of direct and indirect output. Thus, in $1977 this multiplier was 2.25. In $1988 this multiplier was 2.51. In 1988 the agricultural terms of trade were such that the relatively lower agricultural prices deflate the port values of ag products less, resulting in more real output to drive the real multiplier process. When these real final demands are converted, the relatively higher prices in these sectors translate into higher nominal incomes.

For I/O impact analysis at the subnational level regional prices may not be observable and terms of trade may not seem important for small regions. Yet I/O models represent a base year's conditions. The base model locks in a set of relative prices and a fixed relationship between real output and nominal output. Input/output multipliers reflect these underlying relationships. When these relationships change and the I/O analyst ignores the changes, the resulting analysis may be as ridden with money illusion as a consumer spending as though the $33,000 income this year were greater than last year's $30,000 even when inflation is ten percent.

Food and Fiber System

One of the more successful uses of I/O for sectoral analysis at the national level is ERS's use of it to estimate income and employment associated with the Food and Fiber System. Although the series was started less than ten years ago, the Statistical Abstract of the United States has included these estimates for four years. For 1988 we estimate that the Food and Fiber System accounted for $727. billion or 15% of our nation's GNP and jobs for nearly 20 million workers or 16% of the US civilian work force.

The statement that "Agriculture accounts for 16 percent of the U.S. economy," is open to misinterpretation. Many people assume that farming makes up 16 percent--a glaring misconception when, in fact, farming makes up less than 2 percent. Inputs-supplying industries account for another 2 percent and post-farm industries--such as supermarkets, processors, and restaurants--make up the remaining 12 percent.

To understand the size of American agriculture today and its role in the U.S. economy, it helps to examine individual contributions within the agricultural economy, and then to view agriculture as part of the economic whole. To define the role of a sector in the general economy we have to start with the basics. What does a farm sector do? It produces grains, livestock, livestock products, fruits, vegetables, tobacco, cotton, greenhouse and nursery products, and so on. However, this production has no value unless there is a use for it.

Consider for example, the economic role of unsold inventory of a farm machinery manufacturer. Its manufacture has provided jobs and markets for machinery parts and other manufacturing inputs. But now that it is sitting on the manufacturer's lot, it has no further role in the economy until purchased and put to use.

We can view the farm sector similarly. For a vigorously operating economy, production is not a goal, consumption is. Production replaces goods and services consumed by individuals and collectively by private and public institutions. A structural analysis must consider where farm products fit in this view of an economic system which emphasizes the use or consumption of output?

So, the first step in defining the size of the Food and Fiber System is to define the final output of the system. In any year, we define the final output of the system as (1) domestic consumers' expenditures for food, clothing, shoes, tobacco products, flowers, seeds, and potted plants, (2) agricultural exports, (3) the value of farm inventory change, and (4) the value of changes in off-farm private and government stocks of farm commodities.

We adjust these estimates in several ways. Domestic consumer spending includes spending for imported products as well as domestic products. Since we want to characterize the size of the domestic Food and Fiber System, we subtract agricultural and apparel imports from the final product of our system. Ideally, we would like to include as final product only apparel with natural fibers. But, one has only to reflect on the prevalence of manmade-natural fiber blends, the ease of switching fibers in fabric manufacturing, and the vagaries of apparel fashion trends to realize this is not a simple adjustment to make and that an adjustment made one year may be inappropriate the next. We make an adjustment but I don't vouch for its accuracy. Although adjustments are made in the estimation process, its accuracy is questionable.

Once the final output of the Food and Fiber System, is decided this information can be used in an input-output model. This model allows the levels of economic activity in the various sectors of the economy required to support the final output of the Food and Fiber System to be identified. This supporting output can be categorized in three ways: 1) output of raw farm commodities, 2) nonfarm output used to support the production of farm commodities (referred to as backward linkages), and 3) nonfarm output generated in the processing and distribution of raw farm commodities (forward linkages). Most nonfarm industries produce some outputs that support farm production, some that support processing and distribution, and some not used by the Food and Fiber System at all. For example, the Food and Fiber System uses nearly seventy percent of the output of the metal containers industry. About 89 percent of the agriculture-related outputs are forward linkages such as food containers, but 11 percent are backward linkages such as pesticide containers and oil cans.

In addition, because maintaining the capacity to produce farm products requires periodic replacement of and additions to the farm capital stock of machinery, equipment, and structures, farm capital expenditures is included as part of the business activity which supports the Food and Fiber System.

Using this procedure of identifying the sales of the various industries which contribute to the final output of the Food and Fiber System, in 1988 the U.S. Food and Fiber System generated an estimated $727 billion or 15 percent of U.S. gross national product and generated employment for nearly 20.0 million workers or 16 percent of the civilian work force. These percentages appear to be valid for most years, although 1988 was an unusual year. In our example of the unsold farm machinery, manufacturing generates business activity, but as long as equipment sits on the manufacture's lot it does not have any additional effect on the economy. In fact, when buyers do appear their purchases will not stimulate the economy until enough of the machinery stock is sold that the manufacturer chooses to start again producing machines. We can view the Food and Fiber System in 1988 in a similar light. Because of reduced crops in 1988 and two previous years of bumper crops, a higher-than-normal share of 1988 consumption was from previous crops. This, together with the lagged effects of the strong dollar in the mid-80's that dampened exports and encouraged agricultural and apparel imports, led to a smaller estimated role for the Food and Fiber System in the economy in 1988. In 1987 the Food and Fiber System had accounted for 16 percent of GNP and 17 percent of civilian employment.

The use of I/O analysis for constructing estimates of the Food and Fiber System estimates provides several lessons. A major lesson is that the I/O assumption of fixed technical coefficients may be too strong for the farm sector. Weather shocks, technical change, flexible sector prices, and volatile markets combine to introduce variability into farm sector technical coefficients. One way to check the seriousness of this shortcoming involves estimating agricultural value added with a full national income and product account (NIPA). Table 10.1 presents the results of one such set of experiments. At the national level we have the annual gross farm product or gnp originating in farming estimates to use as a standard. A conceptual difference between value-added in the national I/O table and gross farm product involves the treatment of rental income of farm operator landlords and implicit gross rental value of farm dwellings (included in gross farm product, excluded from agricultural value added). Assuming these rental shares hold for all years an I/O- based estimate of gross farm product using a full NIPA should always be 16.8 percent too low (the 1977 experience). Differences from this 16.8 percent warn of the necessity of special consideration of estimated factor income in the farm sector. In fact at the national level, I/O is a rather poor predictor of agricultural factor income. We adjust our Food and Fiber System income estimates by the percentage adjustment implied by the prediction error in the economywide measure of agricultural factor income.

A useful aspect of our I/O -based measure of the Food and Fiber System is the ability to identify the demand source of output, employment, and income generated in the system. Figures 2-4 show these for four broad groupings of demands. Several patterns emerge. Livestock output is essentially domestic consumer demand driven. On balance, the U.S. Food and Fiber System is a domestic demand driven sector. Apparel imports and the strong dollar in the mid-eighties effectively neutralized our strong export position in grains, oilseeds, and cotton. Depending on ones view of the government's role (as reflected in the "other" category) the government is either a source of volatility for the system or an absorber of excess supplies and provider of tight demands. This category provides much of the volatility. And it again reminds us of the farm machinery inventory example. The Food and Fiber System produces outputs for foreign and domestic consumers, both public and private. Without effective demands for this output, stocks accumulate and either prices fall or producers withhold factors, (I/O assumes the latter) to clear the markets.

Economywide Effects of Policy

Nineteen eighty three changed national level agricultural policy analysis. Many agricultural input manufactures had already produced their 1983 supplies before the government announced the PIK program. This program retired nearly 50 million acres and, with them, the demand for inputs on this land. If they had not previously been so aware, this occurrence demonstrated to farm input suppliers just how large a stake they had in the operation of farm programs. One result of this heightened awareness was a more active involvement of farm-related industrial groups in the discussion of the 1985 Farm Bill. A formal expression that they had effectively conveyed this interest to the Congress was the following section from the Conference Report of the Food Security Act of 1985.

> It is the intent of the conferees that whenever the Secretary announces an acreage reduction program, he must send a report to the House and Senate agriculture committees. The report is to contain an economic analysis of the effect of the size of the acreage reduction program on the farm economy - including in particular, the effect of the program on farm income. In addition to the effect of the program on farm income, the analysis is to take into account the effect of the size of the program on farm input and agricultural processing industries, livestock producers, consumers of agricultural products, agricultural trade, jobs,tax revenues, and farm production efficiencies.

In USDA the requirement for an economywide perspective has usually been met with input-output based analysis. Either a standard I/O analysis of the effects of a particular policy on exports or on stocks is used or the Food and Fiber System estimating framework is used to analyze policy shocks for their economywide effects. A particularly successful example would illustrate both the strengths and weaknesses of I/O for analyzing sectoral policy proposals. In the summer of 1986 we responded to an internal USDA request about the effects of the level of farm output on the general economy with an analysis within the FFS framework that suggested small changes in farm output could get absorbed in stock changes and foreign trade adjustments without notably affecting the rest of the economy. It was only when output adjusted enough to influence domestic food consumption that noticeable economywide effects result. Appropriately or not, opponents of supply management programs used this result to question the wisdom of these programs. Proponents of supply management programs modified their proposals. Later impact analyses of supply management programs using economic models with less fixed relationships predict smaller economywide employment impacts. But, these analyses came out after much of the crucial policy action had occurred. The I/O analysis was timely (less than six months from the original request until the publication was on the streets) and was an analysis presented in terms familiar to policymakers. Timeliness and familiarity made it effective. To gain that simplicity and ease of modeling some sacrifice in the reality of substitution possibilities in both production and consumption and an inability to handle price flexibility. In response to these weaknesses , a group in ERS is developing a computable general equilibrium (CGE) modeling capability. But it is going to be a while before a good CGE analyst can produce a analysis of a similar policy question before a good I/O analyst can.

Summary

Having access to data consistent with national income and product accounts data and dealing with a more complete economy where regional leakages cancel make structural analysis at the national level somewhat easier. In the Economic Research Service, input-output analysis is used to do structural analysis in at least three major ways - measuring the impact of agricultural trade, defining the U.S. Food and Fiber System, and measuring the economywide effects of agricultural policy proposals. Or, in the context of this paper's topic - measuring the impacts of a particular sector product, defining the size of that part of the economy linked forward and backward to a particular sector, and measuring the economywide effects of sector specific policies.

These applications have been discussed and several cautionary issues related to I/O analysis have been identified. These procedural considerations include varying relative prices, technical coefficients which may not be as "fixed" as assumed, and overly conservative substitution possibilities.

Although these procedural considerations may complicate the analysis, the benefits of I/O analysis are considerable. I/O allows you to do timely analysis, using an economic model which considers the intersectoral linkages in your communities. Its solution details are consistent with the solution totals. And its results are in a form decision makers can recognize.

References

Edmondson, W. (1989), "U.S. Trade Benefits Economy," *FATUS: Foreign Agricultural Trade of the United States (Sept.-Oct)*, pp. 9-11.

Harrington, D., Schluter, G., and O'Brien, P. (1986), "Agriculture's Links to the National Economy, Income and Employment," AIB-504 (October) 6 pp.

Henry, Mark, and Gerald Schluter, "Measuring Backward and Forward Linkages in the U.S. Food and Fiber System," pp.33-39, Fall 1985.

Lee, Chinkook, Gerald Schluter, and William Edmondson, "Income and Employment Generation in the Food and Fiber System," *Agribusiness*, Vol. 2, No. 2 (1986).

Lee, C., Schluter, G., Edmondson, W., and Wills, D. (1987), "Measuring the Size of U.S. Food and Fiber System", AER-566 (March) 13 pp.

Schluter, G. and Edmondson, W. (1986), "How to Tell How Important Farming is to Your State," *Rural Development Perspectives*, (June) pp. 32-4.

U.S. Bureau of the Census (1989), *Statistical Abstract of the United States: 1989*, (109th Edition), Washington DC.

Table 10.1 Gross Farm Product, Actual and I/O-based Estimates, 1972-1988

Year	Estimated	Actual	Actual/Estimated
1972	35.7	50.3	1.407
1973	39.6	49.5	1.249
1974	37.4	49.3	1.316
1975	41.6	52.6	1.263
1976	40.1	50.7	1.262
1977	43.2	50.4	1.168
1978	46.0	49.5	1.075
1979	47.2	52.4	1.110
1980	41.5	52.0	1.254
1981	50.8	61.4	1.208
1982	43.6	62.4	1.430
1983	40.8	49.8	1.219
1984	50.8	55.5	1.071
1985	43.5	64.4	1.481
1986	43.4	68.2	1.570
1987	45.4	67.6	1.491
1988	45.8	59.0	1.288

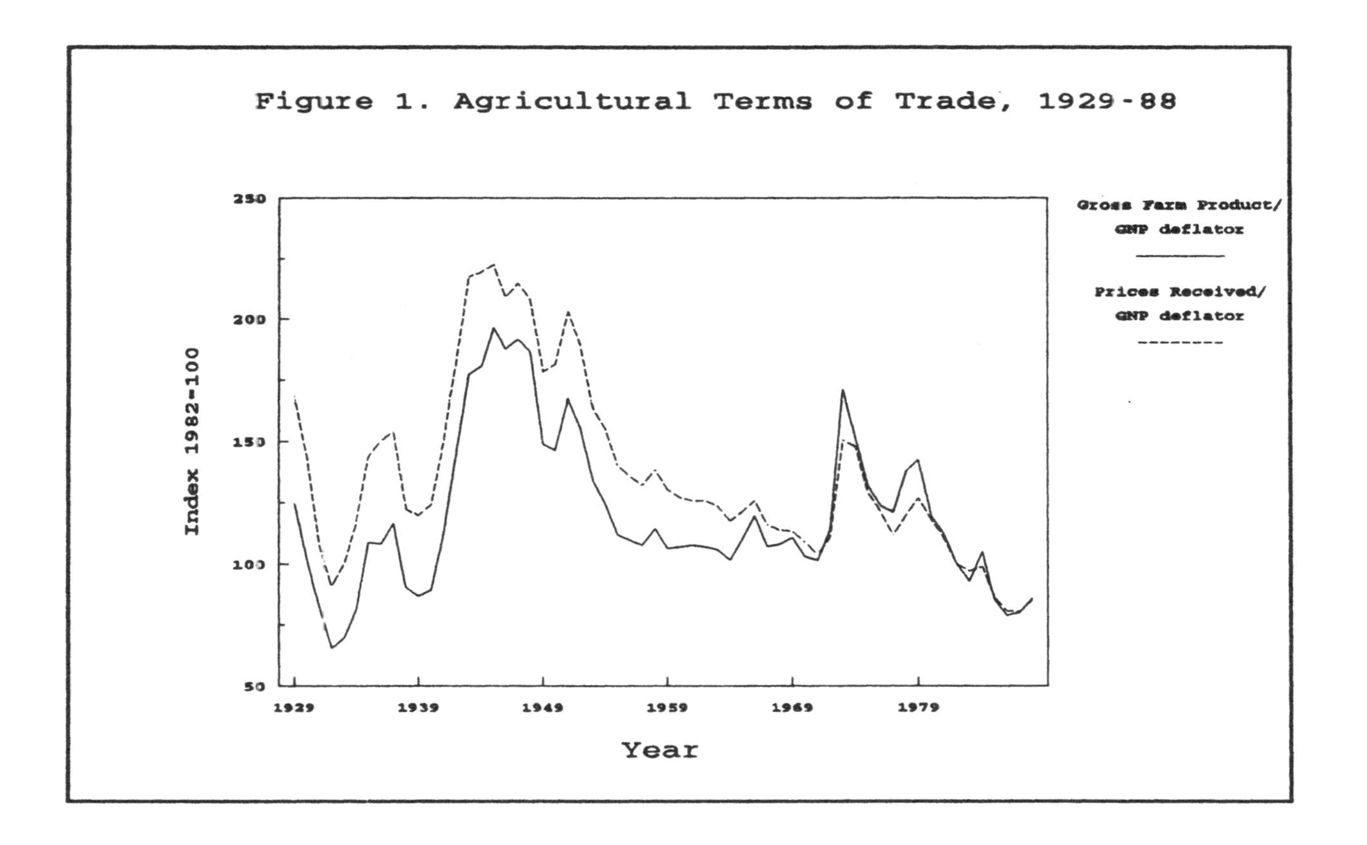
Figure 1. Agricultural Terms of Trade, 1929-88
250
200
150
100
50
Index 1982=100
1929
1939
1949
1959
1969
1979
Year
Gross Farm Product/
GNP deflator
Prices Received/
GNP deflator

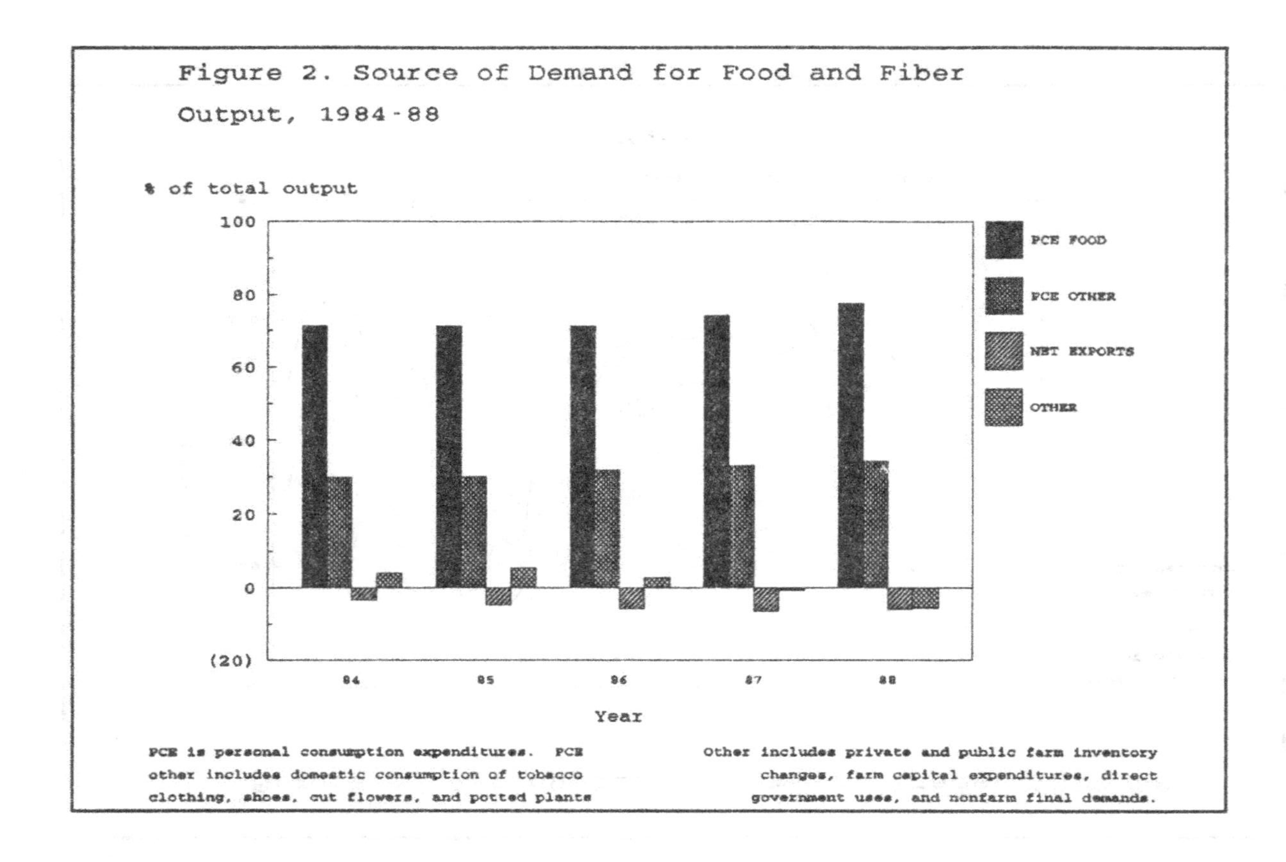

Figure 2. Source of Demand for Food and Fiber Output, 1984-88

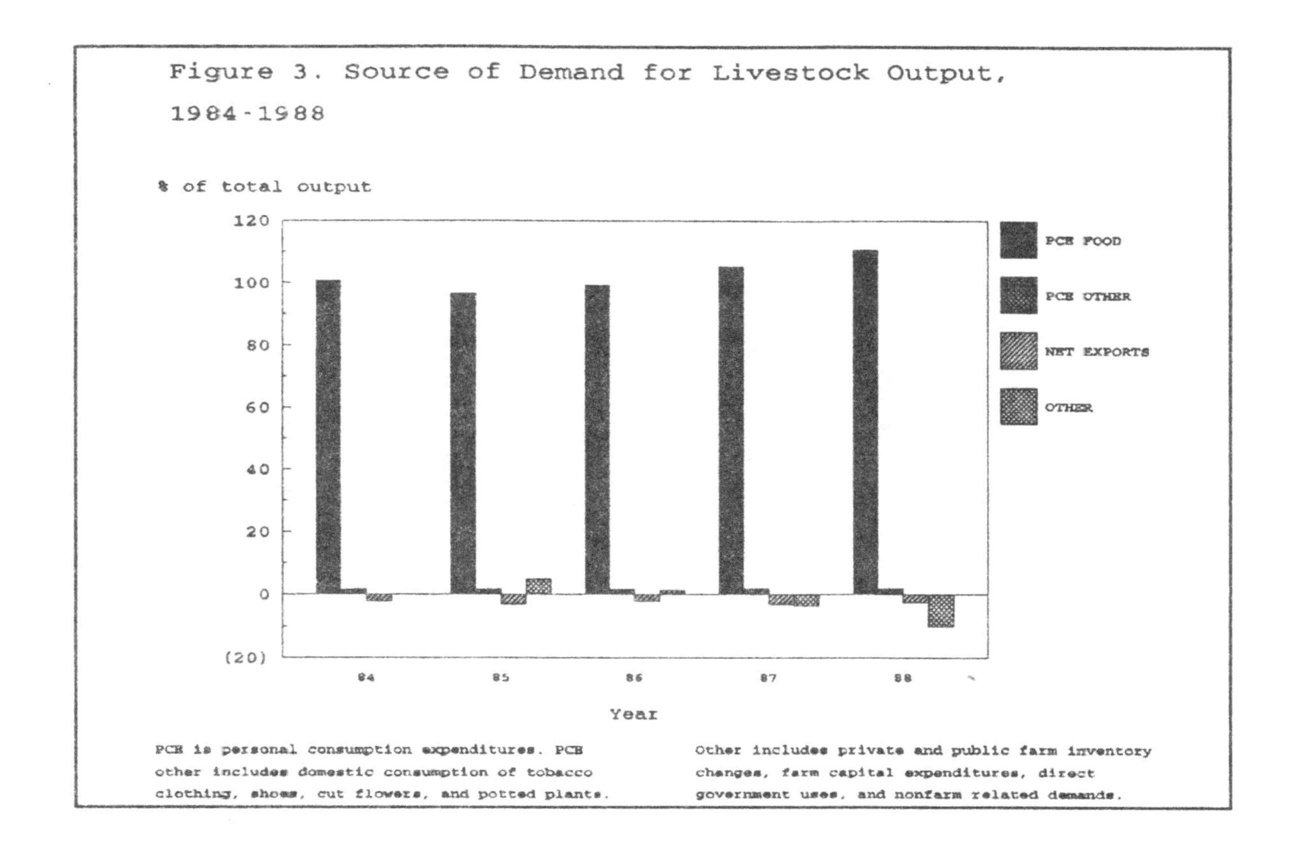

Figure 3. Source of Demand for Livestock Output, 1984-1988

PCE is personal consumption expenditures. PCE other includes domestic consumption of tobacco clothing, shoes, cut flowers, and potted plants.

Other includes private and public farm inventory changes, farm capital expenditures, direct government uses, and nonfarm related demands.

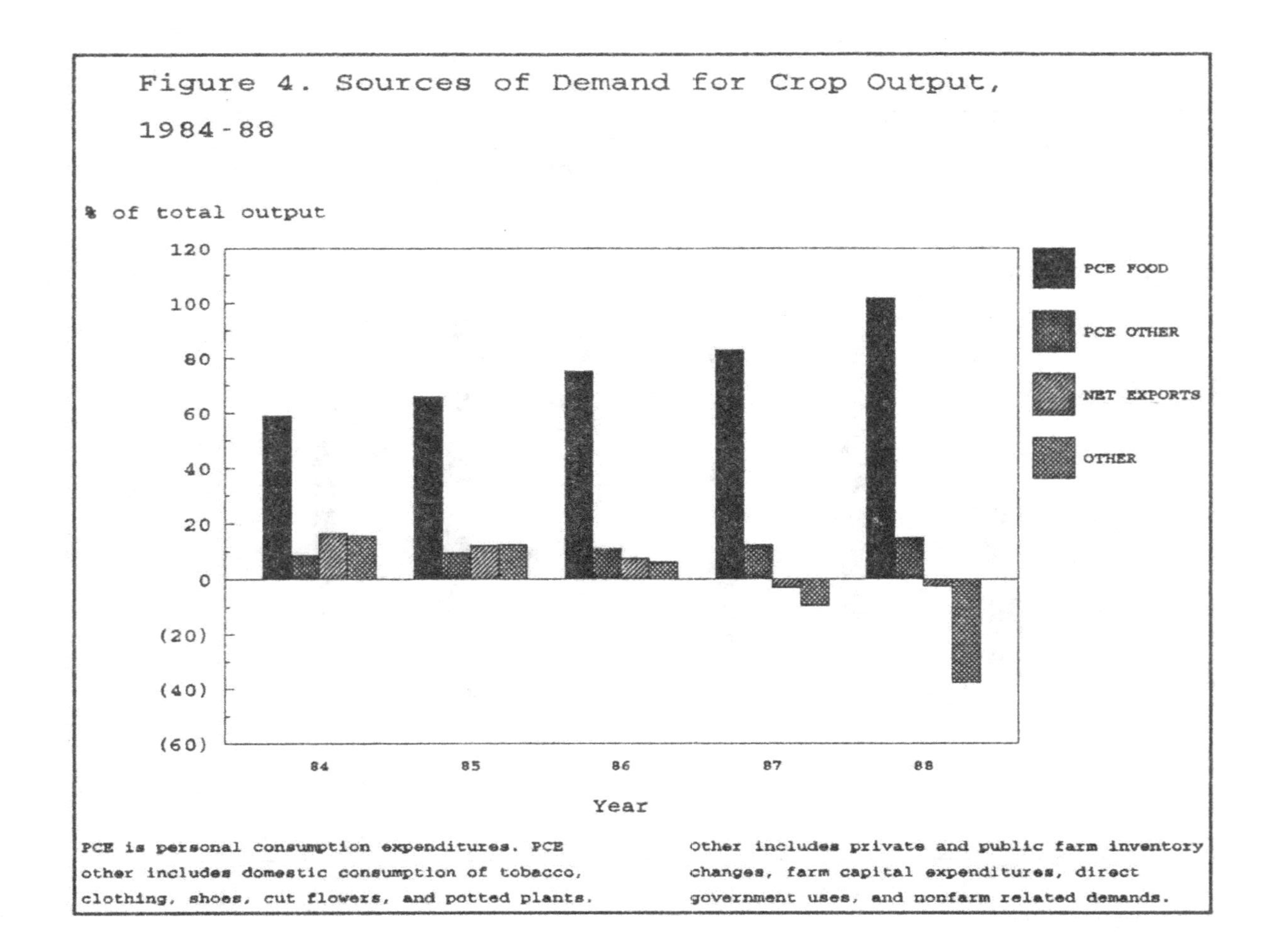
Figure 4. Sources of Demand for Crop Output,
1984-88
% of total output
120
100
80
60
40
20
0
(20)
(40)
(60)
84
85
86
87
88
Year
PCE FOOD
PCE OTHER
NET EXPORTS
OTHER
PCE is personal consumption expenditures. PCE other includes domestic consumption of tobacco, clothing, shoes, cut flowers, and potted plants.
Other includes private and public farm inventory changes, farm capital expenditures, direct government uses, and nonfarm related demands.

11

The Role of Interindustry Linkages in an Industrial Targeting Model

Frank M. Goode

Introduction

The objective of this paper is to describe an industrial targeting model that has been developed and to discuss the important role played by interindustry linkages in the model. The paper is organized as follows: The next section contains a brief discussion of the industrial targeting model. Section III contains a description of the econometric analysis embodied in the targeting model and discusses how interindustry considerations shaped the analysis. Section IV is devoted to a discussion of the construction and empirical implementation of the variables used to investigate the role of interindustry linkages in rural industrial location. A brief discussion of the empirical findings associated with these variables is provided in Section V and the paper is summarized in Section VI.

The Northeast Industrial Targeting and Economic Development Database (NIT and EDD) System

The NIT and EDD system was developed to help rural communities focus their industrial recruitment efforts on those industries that are most likely to locate in their community. Most rural industrial recruitment efforts are very general in nature. For example, communities often develop slick color brochures featuring their communities' attributes and the brochures are mailed to Fortune Five Hundred companies. Presumably, the attributes of a given community are compatible with the requirements of only a few industries. The recruitment efforts are likely to be more successful if they are concentrated on those few industries and emphasize the attributes that are important to those industries. The NIT and EDD system provides the information that makes it possible for communities to concentrate their recruitment efforts. Specifically, the NIT and EDD system provides information that can be used to identify the industries that are most likely to locate in a community and identifies the community attributes that are important to those industries.

Identifying Target Industries

The system contains three indicators that can be used to identify the industries most likely to locate in the community, that is, identify target industries. The first indicator is an estimate of the "probability" of a plant from each of approximately seventy manufacturing industries locating in the community. The probability estimates are referred to as "choice indices." A partial listing of the choice indices for Carbondale, Pennsylvania is shown in Table 11.1. In essence, the choice indices indicate which industries are "best" for Carbondale. This indicator is an important one but others should also be considered.

The industry that is "best" for Carbondale may be even "better" for many other communities. That is, Carbondale may face considerable competition in attracting its "best" industry if many other communities better meet the industry's requirements. Accordingly, the NIT and EDD system provides information regarding how Carbondale's choice index for an industry compares to the choice index for that industry in other communities. This information is shown on Figure 1. Figure 1 is a frequency diagram of the choice indices for the Apparel industry for all small metropolitan communities in the Northeast region. Carbondale's position in the distribution is indicated by the light shaded bar. Thus, Figure 1 indicates that approximately one half of the communities are more likely to attract an Apparel plant than is Carbondale.

Even though the Apparel industry is the second "best" industry for Carbondale, its selection as a target industry is questionable because so many communities have a higher choice index. This is particularly true when the Apparel industry is compared to the Fabricated Textile Products industry. The choice indices in Table 11.1 indicate that the Apparel industry is slightly "better" than the Fabricated Textile Products industry for Carbondale. However, Figure 2 indicates that there are very few communities that have higher choice indices for the Fabricated Textile Products industry than Carbondale. Thus, considering both indicators the Fabricated Textile Products industry is the better target industry for Carbondale.

The third indicator provided by the NIT and EDD system that can be used to select target industries is the "regional share coefficient" for communities that are very similar to Carbondale. A program within NIT and EDD identifies communities that are most similar to Carbondale and calculates the regional share coefficients for those communities. The idea is that if an industry has preformed well in communities similar to Carbondale it may do well in Carbondale and should be a target industry. Appendix Tables 11.1 and 11.2 contain a list of the communities most similar to Carbondale and the regional share coefficients for those communities, respectively.

Identifying Community Attributes

The NIT and EDD system identifies the community attributes that are statistically associated with plant locations in each of the sixty seven manufacturing industries. Table 11.2 contains the attributes that are associated with plant locations in the Construction Equipment industry. The information in this table suggests that plants in the Construction Equipment industry tend to locate in communities that have relatively high incomes, airline service, and colleges. In addition, they prefer communities that have low local and county taxes, well developed service sectors, and are in areas where the oil drilling industry is active. If this industry is a target industry then recruitment information should focus on these attributes.

On the other hand, the attributes associated with the location of Agricultural and Industrial Chemical are completely different. As shown in Table 11.3, the Chemical industry tends to locate in communities where two of its major intermediate manufactured inputs are available and which have a large and underemployed male labor force. In addition, this industry prefers communities that are relatively large and are located near interstate highways. Thus, the community attributes that should be stressed in recruiting an Agricultural and Industrial Chemical plant are completely different from those attributes for a Construction Equipment plant.

In addition to identifying the attributes, the NIT and EDD system contains information on how a community compares to all other communities on a given attribute. For example, local and county taxes were an important attribute for the Construction Equipment industry. Figure 3 indicates how Carbondale compares to all other small metropolitan communities on that attribute. Namely, Carbondale has relatively low local and county taxes.

In summary, the NIT and EDD system provides information that makes it possible for communities to identify a set of target industries. In addition, the system identifies the attributes that are associated with location of plants in those industries. This information is based on an econometric analysis. The following section discusses this analysis and indicates how interindustry considerations played a dominant role in shaping the analysis.

Estimating Choice Indices

General Methodology

The choice indices are the foundation of the NIT and EDD system. The choice indices were estimated by using the following econometric model:

$$(1) \quad {}_{i}PL_{70\text{-}80} = a + b_1\,{}_{70}CA_{1,i} + b_2\,{}_{70}CA_{2,1} + \text{--} + b_{30}\,CA_{30,i} + e_i$$

Where:

${}_{i}PL_{70\text{-}80}$ = 1 if a plant located in community i during the period 1970-1980, zero otherwise.

${}_{70}CA_{1,i}$ - ${}_{70}CA_{30,i}$ = is the value of thirty community attributes in 1970 for community i.

a and b_1 - b_{30} = parameters to be estimated.

e_i = error term.

The general hypothesis embodied in this model is that plant location in a community during the 1970-1980 period is a function of the community attributes in 1970. After the above model was estimated the following formula was used to estimate the choice indices:

$$(2) \quad {}_{i}CI_{80\text{-}90} = \hat{a} + \hat{b}_1\,{}_{80}CA_{1,i} + \hat{b}_2\,{}_{80}CA_{2,1} + \text{--} + \hat{b}_{30}\,{}_{80}CA_{30,i}$$

Where:

${}_{i}CI_{80-90}$ = the choice index for community i for the 1980-1990 period.

${}_{80}CA_{1,I}$ - ${}_{80}CA_{30,i}$ = the value of thirty community attributes in 1980 for community i.

$\hat{a}$ and $\hat{b}_1$ -$\hat{b}_{30}$ = estimated parameters.

In essence, the choice indices were obtained by estimating a "structural equation" which represented the relationship between plant location during the 1970-1980 period and the community attributes in 1970. The values of the community attributes for 1980 were substituted into the "structural equation" to generate estimates of the probability of the community being the site of a plant location during the 1980's, that is, a choice index. In the next section we discuss in greater detail the role of interindustry linkages in the formulation of the econometric model.

Incorporating Interindustry Linkages into the Econometric Model and the Lack of Interindustry Variables in Previous Studies

Many studies have been conducted focusing on rural industrial location. These studies have almost completely ignored the role that interindustry linkages play in the location decision of firms. The explanatory variables typically used in these studies reflect general social and economic characteristics such as per capita income, labor force participation, public service availability, and community amenities. Frequently, the studies include variables reflecting proximity to colleges, highways, and rail and air service. The only explanatory variables used in these studies that in any sense relate to interindustry linkages are those variables used to reflect agglomeration and market effects.

The typical variable used to reflect agglomeration effects is "the proportion of a community's employment that is in the manufacturing sector." This variable is at best a very crude measure of interindustry linkage. Interindustry linkages refer to the specific intermediate manufactured inputs used by an industry, not a general measure of the relative importance of the manufacturing sector in a community's economy. In many instances a rural community's economic base is dominated by a single large firm and this variable does not adequately reflect either agglomeration economies or industrial linkages. Perhaps a more important shortcoming of this variable is that it does not consider spatial effects. That is, rural communities are so small that there are unlikely to be appreciable agglomeration economies or industrial linkages within the community, but there may be industrial linkages among plants in adjacent communities.

Similarly, these studies have used only crude measures of industrial linkages on the output side. The typical variable used in rural location studies to reflect market access is "distance to the nearest Metropolitan Statistical Area." This variable is based on the assumption that the output of rural manufactures is destined for urban areas either as final consumer products or as intermediate inputs for other manufactures. The decentralization of the manufacturing industries in the past fifty years makes this assumption difficult to justify. Given the rather feeble attempts to investigate the role of interindustry linkages in industrial location decisions it was decided that the location analysis incorporated in the NIT and EDD system would emphasized these linkages. The next two sections describe the general principles involved in construction, the interindustry variables, and the data used to empirically implement them.

General Principals Involved in the Construction of the Interindustry Variables

To test hypotheses concerning the role of interindustry linkages in industrial location decisions it was necessary to construct variables that reflect the availability of intermediate inputs and market access. If these variables are significantly associated with the location of new plants then it is possible to conclude that interindustry linkages are important factors in the location decision. To test this hypothesis it is necessary that the statistical analysis utilize detailed industrial classes. That is, the interindustry linkages for seemingly similar industries are very different. For example, the Standard Industrial Classification (SIC) industry 2511 is the Wood household furniture, except upholstered industry. The industry most similar to SIC 2511 is SIC 2512 and the latter is the Wood household furniture, upholstered industry. These seemingly similar industries have very different industrial linkages. Both industries are of course linked to the primary wood manufacturing industry. However, the upholstered furniture industry has important links to the plastics and the broadwoven fabric industries. The non-upholstered industry does not have these linkages. If these two seemingly similar industries were aggregated, the effects of interindustry linkages would likely be "washed out" in the aggregation process. Accordingly, this study investigates the factors associated with the location of sixty nine manufacturing industries. That is, all four digit SIC manufacturing industries were aggregated into sixty nine "aggregate industries." The aggregate industries are listed in appendix Table 11.5. Only industries with similar "input structures" were aggregated.

In addition to constructing the variables for detailed industrial classes, the variables had to have a spatial dimension larger than a community. That is, the unit of analysis used in this study was "communities." Most rural communities have such small industrial sectors that interindustry linkages within the community are unlikely. However, interindustry linkage among rural communities in the same proximity are possible. Accordingly, the spatial dimension of the linkage variables was an area with a radius of two hundred miles centered on the community.

Clearly, the potential for interindustry linkages is greater if the distance between linked industries is small. That is, even though the spatial dimension of the variable was two hundred miles, a plant located two hundred miles from the community is less likely to be linked than a plant two miles away. Accordingly, the linkage variables discounted the impact of a distant plant relative to a nearby plant. The discount factor was based on the distance to the plant and the cost of transporting the product involved.

The final general principle involved in the construction of the linkage variables was that the variables reflect net rather than gross input supply. If an industry has strong interindustry linkages it will seek a site in an area where the inputs purchased from other industries are in excess supply. There may be areas where large quantities of the input are produced, but if there is an excess demand for the input, a new plant will be unlikely to locate in the area because of relatively high input price and the likely need to import the input from outside the area. Thus, plants are more likely to locate in an area with large excess quantity supplied, not necessarily in an area with a large quantity supplied. Thus, the linkage variables reflect excess or net supply.

Variable Definition and Empirical Implementation

Input variables were constructed for each of the three most important intermediate inputs used by an industry. These inputs are listed in Appendix Table 11.5. The three important inputs were defined as those with the largest direct requirements coefficients in the national input-output study. A single market variable was constructed for each of the industries.

The Potential Net Input Availability Variable

The input variables are referred to as potential net input availability (PNIA) and are defined as:

$$(3) \quad PNIA_j = \sum_{i=1}^{n} \frac{P_i - C_i}{T_{ij}}$$

Where:

$PNIA_j$ is the potential net input availability in community j.

n is the number of communities within two hundred miles of community j.

P_i is the production of the input in community i.

C_i is the consumption of the input in community i.

T_{ij} is the cost of transporting the input from community i to community j.

In essence, this variable reflects whether community j is located in an area with a surplus or deficit supply of the input. For a more detailed discussion of the variable see Hastings (Hastings, 1982).

The PNIA variable where empirically implemented follows:

Steps Used to Estimate Total Production

Step 1. Output per employee for each industry was estimated using data from the DUNS Market Indicators File.

Step 2. Employment levels in each industry in each community were obtained from the DUNS Market Indicators File.

Step 3. The total production in an industry in a community (P_i) was obtained by multiplying a community's employment in an industry by the industry's output per employee.

Steps Used to Estimate Total Consumption

Step 1. Total output of each industry in each community was estimated as described above.

Step 2. The consumption of an input by an industry was estimated by multiplying the total output of the industry in a community by the direct requirements coefficient for the industry as reported in the national input-output study.

Step 3. The total consumption of an input in a community (C_i) was obtained by summing the consumption of all industries (and final demand sectors) in the community.

Steps Used to Estimate Transportation Costs (T_{ij})

Step 1. Data were obtained from regional trucking associations on the cost of transporting various commodities between communities in the Northeast.

Step 2. The distance (D_{ij}) between all communities in the Northeast was obtained from road maps.

Step 3. A regression model was estimated for each commodity. The dependent variable was transportation cost and the independent variable was the distance between the two communities.

Step 4. Each T_{ij} was estimated using the following formula:

$$(4)\quad T_{ij} = \hat{a} + \hat{b} D_{ij}$$

The Market Access Variable

The market access variable is defined as follows:

$$(5) \quad MA_j = \sum_{i=1}^{n} \frac{C_i - P_j}{T_{ij}}$$

Where: MA_j is the market access for firms in community j. n, C_i, P_i, and T_j are as defined for the PNIA variable.

This variable reflects whether or not community j is in a region where there is an excess demand for the product. When this variable is positive there is an excess demand. For a more detailed discussion of this variable see Goode (Goode, 1986). There are two hypotheses concerning the sign of the coefficients for the variable. If the plants in an industry respond to agglomeration effects and concentrate spatially, then the signs of the coefficients are expected to be negative. That is, if the plants concentrate they will be expected to locate in an area that has an excess supply (negative excess demand). If the plants tend to disperse spatially, each having its market area, then plants would be expected to locate where there is an excess demand.

The market access variable involves the same components that are used to implement the input variables and the procedures described above were used to implement the market variable. The following section contains a discussion of the empirical findings.

Empirical Findings

Market Access

Table 11.4 contains a summary of the empirical findings. The findings associated with the market access variable were the most surprising and perhaps the most important. For decades the conventional explanation for the concentration of manufacturing activity in urban areas was agglomeration economies. In the 1970's the observed pattern of manufacturing employment decentralizing into rural areas was presumed to be due to agglomeration economies becoming less important. The findings of this study are consistent with the hypothesis that agglomeration economies are important in explaining rural industrial location. The market access variable was significantly different from zero for 35 of the 67 industries involved in the study. However, of the 35 significant coefficients 29 carried a negative sign. These results indicate that plants in many of the industries are choosing locations in areas where there is a surplus of the product they produce. Again, these results are consistent with the hypothesis that plants tend to cluster near one another in order to take advantage of agglomeration economies.

The findings are also consistent with other hypotheses. For example, the findings are consistent with what could be called the "copycat" location decision. It can be argued that most plants locating in rural communities are small and do not have the resources to devote to intensive location search processes. As a result, entrepreneurs planning to start a new plant may look to establish in an area where similar businesses are doing well.

Whatever the reason, the findings clearly suggest that plants locating in rural communities tend to locate in areas where there are numerous existing plants. This finding is important for predicting the location of new plants and for a more general understanding of the rural industrial location process.

Intermediate Input Availability

The findings associated with the input variables were not as conclusive as findings relating to the market variable. There were 201 input variables in the 67 regression models. Of the 201 variables only 53 had coefficients that were statistically different from zero. Of the 53 variables 24 carried negative signs. The negative signs on the input variables are difficult to rationalize in terms of economic theory. A more probable explanation involves the structure of industries combined with the process of classifying industries.

For example, the findings in Table 11.4 indicate that the location of plants in the Sawmill industry (SIC 2421) are negatively associated with the availability of inputs from the Logging industry (SIC 2411). This finding is inconsistent with economic theory and causal observation. The Sawmill industry uses a heavy and bulky input and theory clearly suggests that plants from this industry will locate near supplies of that input. Casual observation also suggests that the Sawmill industry and the Logging industry are closely linked. In fact they are so closely linked that they are almost always a part of the same firm. When these firms are classified according to their major output they are nearly always classified as Sawmills. Accordingly, due to classification processes, areas with several sawmills show little employment in the Logging industry, resulting in a negative correlation between employment in the Logging and Sawmill industries. (Note that the Logging and Sawmill industries are in the same 2 digit SIC class, namely, SIC 24.)

Of the 24 negative coefficients associated with the input variables, 19 involved cases in which the input was in the same 2 digit industry or was associated with the paper containers input variable. Thus, it appears that the combination of industrial structure and the industrial classification system leads to negative correlations between location of plants and the availability of inputs from closely linked industries.

On the other hand, when the input industry is not in the same 2 digit class, the findings of this study are consistent with the hypothesis that the location of selected industries is associated with the interindustry linkage hypothesis. For example, the location of plants in the Leather industry (SIC 31) are positively associate with the inputs provided by the Meat Packing industry (SIC 20).

Summary

This paper reports on a study designed to test the hypothesis that interindustry linkages are important determinants of the location of manufacturing plants. The findings suggest that market linkages are important for a few industries but the more important finding is that plants in an industry tend to cluster geographically. This clustering may be the results of agglomeration economies, "copycat" search procedures, or other factors.

Despite data source problems that muddy the findings with regard to input linkages, it does appear that firms in selected industries do tend to locate in areas where there is an excess supply of the intermediate manufactured inputs they use. Since this is the first study to investigate the relationship between industrial linkages and industrial location it is not surprising that unanticipated data problems were encountered. Accordingly, the precise role of interindustry linkages in industrial location decisions await additional research.

References

Goode, Frank, "The Efficacy of More Refined Demand Variables." *Growth and Change*, Vol. 17, No. 1, January 1986, pp. 66-75.

Hastings, Steven E., and Frank M. Goode, "An Input Supply Approach: Improved Measures of Industrial Location Factors." *Growth and Change* Vol. 13, No. 0, July 1982, pp. 25-31.

Table 11.1 Choice Indices for Carbondale, Pennsylvania

Choice	Index	Industry Group
1	0.63	Plastics
2	0.59	Apparel
3	0.54	Fabricated Textile Products
4	0.53	Construction Equipment
5	0.50	Metal Household Furniture
	0.62	Plastics
	0.58	Apparel
	0.53	Fabricated Textile Products
	0.52	Construction Equipment
	0.49	Metal Household Furniture
	0.45	Wood Furniture
	0.44	Wood Veneer and Containers
	0.43	Structural Metal
	0.42	Mobile Homes and Campers
	0.40	Aluminum Foundries
	0.40	Narrow Fabric
	0.39	Fish and Canned Specialties
	0.39	Leather Products
	0.38	Other Concrete Products
	0.38	Coating and Engraving

Table 11.2 Community Attributes for the Construction Equipment Industry

52 Construction Equipment	
Regression Coefficient	Variable Name
0.0793	Per Capita Income
0.0354	Number of Airlines Serving the Community
0.2776	Proportion of Population Enrolled As College Undergrad
-0.0007	Local + County Tax Per Capita
-0.0874	If Service Structure is Basic Rather Than Complex
2.5387	Input Availability (IA)-Petroleum Drilling

Table 11.3 Community Attributes for the Agricultural and Industrial Chemical

27 Agricultural and Industrial Chemical	
Regression Coefficient	Variable Name
5.9145	Input Availability (IA)-Paper Containers
4.8749	Input Availability (IA)-Plastics
-1.2072	Proportion of Population Over 64 Years Old
-0.9459	Proportion of Males who Worked At Least 27 Weeks in 1979
0.0027	Number of Miles to Interstate Highway
0.0036	Community Population

Table 11.4 Summary of Regression Results for Nonmetropolitan Communities*

Industry	Input #1	Input #2	Input #3	Market
Meat Packing Plants	.	.	.	.
Poultry and Dairy Processing	N	.	.	.
Fish and Canned Specialties	.	P	P	N
Grain and Seed Processing	.	.	N	N
Miscellaneous Foods	.	.	.	P
Tobacco	.	.	.	.
Broadwoven Fabric	.	.	N	.
Narrow fabric	N	.	N	N
Miscellaneous Textiles	.	.	.	N
Fabricated Textile Products	P	.	P	.
Logging	.	.	.	P
Sawmills	N	.	.	.
Hardwood Flooring	.	.	.	N
Specialized Sawmills	.	.	.	N
Wood Furniture	.	N	.	N
Wood Veneer and Containers	.	N	.	N
Miscellaneous Wood Products	N	P	.	N
Metal Household Furniture	.	.	.	.
Metal Office Furniture	N	.	P	.
Paper Mills	N	.	P	.
Paper Containers	.	.	.	.
Newspaper Publishing	P	.	.	.

(continued)

Table 11.4 Summary of Regression Results for Nonmetropolitan Communities

Periodical Publishing	P	.	P	.
Book Publishing	N	N	.	N
Binding and Card Printing	.	.	N	N
Commercial Printing	.	.	.	.
Agricultural and Industrial Chemicals	P	.	.	N
Plastics	N	.	.	.
Drugs	.	.	.	N
Refined Petroleum	.	.	.	P
Tire	P	.	.	P
Leather Products	P	.	N	N
Glass and Glassware	.	.	.	N
Stone and Clay Products	P	.	.	.
Concrete Block and Brick	.	.	.	.
Other Concrete Products	P	.	N	.
Ready Mix Concrete	.	.	.	N
Gray Iron Foundries	.	.	P	.
Lead Smelting	.	.	.	.
Zinc Smelting	P	.	P	.
Aluminum Foundries	.	.	P	.
Metal Bolts and Wire	.	.	.	.
Nonferrous Smelting	P	.	.	.
Copper Wire	.	.	.	N
Cutlery and Hand Tools	.	.	.	N
Heating and Plumbing Fixtures	.	.	P	N
Structural Metal	.	.	.	N
Coating and Engraving	P	.	.	P
Miscellaneous Metal Products	.	.	.	N
Engines	.	P	.	.
Mobile Homes and Campers	.	N	.	N
Construction Equipment	.	P	.	.
Machine Tools and Dies	.	.	.	N
Special Industrial Equipment	N	N	.	N
General Industrial Equipment	.	P	.	.
Service Machines	.	N	.	.
Radio, TV, Telephones	.	.	N	.
Household Appliances	.	.	N	.
Electronic Components	.	.	.	N
Motor Vehicles	.	.	.	P
Measuring Instruments	.	.	.	.

(continued)

Table 11.4 Summary of Regression Results for Nonmetropolitan Communities

Jewelry and Musical Instruments	.	.	P	N
Burial Caskets	.	P	.	N
Sporting Goods	.	.	.	.
Advertising Signs	.	.	N	N
Miscellaneous Manufacturing	.	.	P	N
Apparel . .	.	.		

*"P" indicates a position coefficient that was significantly different from zero at the 10% level. "N" indicates a significant negative coefficient.

Appendix Table 11.1 Comparable Communities for Carbondale, Pennsylvania

The Five Communities Most Comparable to Carbondale, PA
For Each of the Five Selected Industries

Plastics
1 = Riverhead (U) NY
2 = McClure PA
3 = Corry PA
4 = Sommerville MA
5 = Lake Hiawatha (U) NJ

Apparel
1 = Moravia NY
2 = Dannemora NY
3 = Gassaway WV
4 = Farmington Center (U) ME
5 = Franklin NJ

Fabricated Textile Products
1 = Frenchtown NJ
2 = Rutland Center (U) MA
3 = Palmyra NY
4 = Merriewold Lake (U) NY
5 = Staunton VA

Construction Equipment
1 = Canajoharie NY
2 = Freehold NY
3 = Culpeper VA
4 = Egg Harbor City NJ
5 = Woodbine NJ

Metal Household Furniture
1 = Attleboro MA
2 = Princeton WV
3 = Houtzdale PA
4 = Lebanon PA
5 = Mexico NY

Appendix Table 11.2 Regional Share Coefficients for Comparable Communities

Macro Level = State	Carbondale, PA				
	REGIONAL SHARE TABLE				
Industry	Community Number[2]				
	1	2	3	4	5
Plastics	15.8	-0.8	109.8	93.5	533.8
Apparel	-57.8	0.0	0.0	0.0	60.6
Fabricated Textile Products	0.0	0.0	0.0	0.0	0.0
Construction Equipment	0.0	0.0	0.0	0.0	0.0
Metal Household Furniture	0.0	0.0	0.0	0.0	0.0

[12] Community Numbers are Relative to Industry (See Comparables for Names)

Appendix Table 11.3 Independent Variables Used in the Industrial Location Model

Variable Name	Carbondale, PA
61 Input Availability (IA)-Ship and Rail Building	
62 Input Availability (IA)-Measuring Instruments	
63 Input Availability (IA)-Jewelry & Musical Instrum.	
64 Input Availability (IA)-Burial Caskets	
65 Input Availability (IA)-Sporting Goods	
66 Input Availability (IA)-Advertising Signs	
67 Input Availability (IA)-Miscellaneous Manufacturing	
68 Input Availability (IA)-Apparel	
69 Input Availability (IA)-Military Tanks	
70 Whether Community was SMSA or non-SMSA in 1970	
71 Proportion of Population That is Nonwhite	
72 Proportion of Population Over 64 Years Old	
73 Male Labor Force Participation Rate	
74 Female Labor Force Participation Rate	
75 Proportion of Males Who Worked at Least 27 Weeks in 1979	
76 Proportion of Females Who Worked at Least 27 Weeks in 1979	
77 Per Capita Income	
78 Number of Miles to Nearest SMSA	
79 Proportion of Year-round Housing Units on Public Sewer System	
80 Proportion of Population Employed in Manufacturing	

Appendix Table 11.4 Independent Variables Used in the Industrial Location Model

Variable Name	Carbondale, PA
81 Proportion of Population Living in the Same County as in 1975	
82 Proportion of Population Less Than 18 Years Old	
83 Proportion of Population Over 25 Years With College Degree	
84 Number of Railroads Serving the Community	
85 Number of Airlines Serving the Community	
86 Number of Miles to Interstate Highway	
87 Number of Miles to Primary Road	
88 Number of Hospital Beds per 100 Persons	
89 Proportion of Year-round Housing Units That Are Vacant	
90 Community Population	
91 If Community in New England Rather Than South Atlantic	
92 If Community in Midatlantic Rather Than South Atlantic	
93 Proportion of Population Enrolled As College Undergrad	
94 Local + County Tax Per Capita	
95 State Tax Per Capita	
96 If Service Structure is Moderate Rather Than Complex	
97 If Service Structure is Basic Rather Than Complex	
98 Input Availability (IA)-Forestry	
99 Input Availability (IA)-Energy Exploration	
100 Input Availability (IA)-Iron Mining	

Appendix Table 11.5 Industries Included in the NIT and EDD System.

No. Name	1977 SIC's in Industry	Input #	Input Industries
1 Meat Packing Plants	2011 2013 2032 2047	1	Miscellaneous Foods
	2048	2	Poultry and Dairy Processing
		3	Metal Bolts and Wire
2 Poultry and Dairy	2016 2017 2021 2022	1	Paper Containers
	2023 2024 2026 2041	2	Grain and Seed Processing
	2041 2044 2046	3	Metal Bolts and Wire
3 Fish and Canned Specialties	2091 2092	1	Paper Containers
		2	Metal Bolts and Wire
		3	Commercial Printing

(continued)

Appendix Table 11.5 Industries Included in the NIT and EDD System

	Industry	SIC Codes	Rank	Linked Industry
4	Grain and Seed Processing	2033 2034 2035 2037 2038 2043 2045 2051 2052 2061 2062 2063 2065 2066 2067 2082 2083 2084 2085 2086 2087	1	Metal Bolts and Wire
			2	Poultry and Dairy Processing
			3	Paper Containers
5	Miscellaneous Foods	2074 2075 2076 2077 2079 2095 2097 2098 2099	1	Paper Containers
			2	Metal Bolts and Wire
			3	Plastics
6	Tobacco	2111 2121 2131 2141	1	Plastics
			2	Paper Containers
			3	Paper Mills
7	Broadwoven Fabric	2211 2221 2231 2261 2262	1	Ag and Industrial Chemicals
			2	Miscellaneous Textiles
			3	Plastics
8	Narrow Fabric	2241 2251 2252 2253 2254 2257 2258 2259	1	Miscellaneous Textiles
			2	Ag and Industrial Chemicals
			3	Paper Containers
9	Miscellaneous Textiles	2269 2271 2272 2279 2281 2282 2283 2284 2291 2292 2293 2294 2295 2296 2297 2298 2299	1	Ag and Industrial Chemicals
			2	Plastics
			3	Broadwoven Fabric
10	Fabricated Textile Products	2391 2392 2393 2394 2395 2396 2397 2399	1	Broadwoven Fabric
			2	Miscellaneous Textiles
			3	Plastics
11.	Logging	2411	1	Metal Bolts and Wire
			2	Refined Petroleum
			3	Cutlery and Hand Tools
12	Sawmills	2421	1	Logging
			2	Refined Petroleum
			3	Miscellaneous Metal Products
13	Hardwood Flooring	2426 2512	1	Broadwoven Fabric
			2	Sawmills
			3	Plastics
14	Specialized Sawmills	2429 2491	1	Logging
			2	Sawmills
			3	Ag and Industrial Chemicals

(continued)

Appendix Table 11.5 Industries Included in the NIT and EDD System

15	Wood Furniture	2431 2434 2511 2517	1	Sawmills
			2	Wood Veneer and Containers
			3	Cutlery and Hand Tools
16	Wood Veneer and Containers	2435 2436 2439 2441 2449 2452	1	Logging
			2	Sawmills
			3	Plastics
17	Miscellaneous Wood Products	2448 2492 2499	1	Sawmills
			2	Plastics
			3	Logging
18	Metal Household Furniture	2514 2519	1	Plastics
			2	Gray Iron Foundries
			3	Aluminum Foundries
19	Metal Office Furniture	2515 2521 2522 2531 2541 2542 2591 2599	1	Gray Iron Foundries
			2	Plastics
			3	Metal Bolts and Wire
20	Paper Mills	2611 2621 2631 2661	1	Logging
			2	Sawmills
			3	Ag and Industrial Chemicals
21	Paper Containers	2641 2642 2643 2645 2646 2647 2648 2649 2651 2652 2653 2654 2655	1	Paper Mills
			2	Plastics
			3	Ag and Industrial Chemicals
22	Newspaper Publishing	2711	1	Paper Mills
			2	Commercial Printing
			3	Measuring Instruments
23	Periodical Publishing	2721 2741	1	Commercial Printing
			2	Paper Mills
			3	Book Publishing
24	Book Publishing	2731 2732	1	Paper Mills
			2	Binding and Card Printing
			3	Plastics
25	Binding and Card Printing	2761 2771 2782 2789 2791 2793 2794 2799	1	Paper Mills
			2	Plastics
			3	Paper Containers
26	Commercial Printing	2751 2752 2753 2754 2795	1	Paper Mills
			2	Binding and Card Printing
			3	Ag and Industrial Chemicals

(continued)

Appendix Table 11.5 Industries Included in the NIT and EDD System

	Industry	SIC Codes		Linked Industry
27	Ag and Industrial Chemicals	2812 2813 2816 2822	1	Plastics
		2823 2824 2841 2843	2	Paper Containers
		2844 2861 2865 2875	3	Metal Bolts and Wire
		2892 2893 2895		
28	Plastics	2821 2851 2891 2899	1	Ag and Industrial Chemicals 2982
		3021 3041 3069	2	Paper Containers
		3079 3199 3292 3293	3	Metal Bolts and Wire
		3592 3599		
29	Drugs	2831 2832 2833 2834	1	Ag and Industrial Chemicals
			2	Plastics
			3	Paper Containers
30	Refined Petroleum	2911 2951 2992 2999	1	Ag and Industrial Chemicals
			2	Metal Bolts and Wire
			3	Plastics
31	Tires	3011 3031	1	Ag and Industrial Chemicals
			2	Miscellaneous Textiles
			3	Plastics
32	Leather Products	3111 3131 3142 3143	1	Meat Packing Plants
		3144 3149 3151 3161	2	Plastics
		3171 3172	3	Miscellaneous Textiles
33	Glass and Glassware	3211 3229 3231	1	Plastics
			2	Paper Containers
			3	Ag and Industrial Chemicals
34	Stone and Clay Products	3241 3251 3253 3255	1	Plastics
		3259 3261 3262 3263	2	Ag and Industrial Chemicals
		3264 3269 3274 3275	3	Paper Containers
		3281 3291 3295 3296		
		3297 3299		
35	Concrete Block and Brick	3271	1	Stone and Clay Products
			2	Refined Petroleum
			3	Lead Smelting
36	Other Concrete Products	3272	1	Stone and Clay Products
			2	Metal Bolts and Wire
			3	Lead Smelting
37	Ready Mix Concrete	3273	1	Stone and Clay Products
			2	Refined Petroleum
			3	Plastics

(continued)

Appendix Table 11.5 Industries Included in the NIT and EDD System

38 Gray Iron Foundries	3312 3313 3315 3316	1	General Industrial Equipment
	3317 3321 3322 3324	2	Ag and Industrial Chemicals
	3325 3462	3	Zinc Smelting
39 Lead Smelting	3221 3332	1	Paper Containers
		2	Ag and Industrial Chemicals
		3	Stone and Clay Products
40 Zinc Smelting	3331 3333 3339	1	Refined Petroleum
		2	Gray Iron Foundries
		3	Ag and Industrial Chemicals
41 Aluminum Foundries	3334 3353 3354 3355	1	Zinc Smelting
	3361 3369 3398 3399	2	Ag and Industrial Chemicals
	3463	3	Plastics
42 Metal Bolts and Wire	3411 3412 3451 3452	1	Gray Iron Foundries
	3465 3466 3469 3482	2	Aluminum Foundries
	3484 3489 3493 3495	3	Plastics
	3496		
43 Nonferrous Smelting	3341 3356	1	Zinc Smelting
		2	Ag and Industrial Chemicals
		3	Plastics
44 Copper Wire	3351 3357 3362	1	Zinc Smelting
		2	Aluminum Foundries
		3	Plastics
45 Cutlery and Hand Tools	3421 3423 3425 3429	1	Gray Iron Foundries
		2	Plastics
		3	Metal Bolts and Wire
46 Heating and Plumbing Fixtures	3431 3432 3433 3585	1	Radio, TV, Telephones
		2	Gray Iron Foundries
		3	Copper Wire
47 Structural Metal	3441 3442 3443 3444	1	Gray Iron Foundries
	3446 3448 3449	2	Aluminum Foundries
		3	Metal Bolts and Wire
48 Coating and Engraving	3471 3479	1	Plastics
		2	Ag and Industrial Chemicals
		3	Zinc Smelting
49 Miscellaneous Metal Products	3493 3497 3498 3499	1	Gray Iron
		2	Aluminum Foundries
		3	Plastics

(continued)

Appendix Table 11.5 Industries Included in the NIT and EDD Systems

50 Engines	3511 3519	1	Gray Iron Foundries
		2	Plastics
		3	Aluminum Foundries
51 Mobile Homes and Campers	2451 3523 3524 3716 3751 3792 3799	1	Gray Iron Foundries
		2	Motor Vehicles
		3	Heating and Plumbing Fixtures
52 Construction Equipment	3531 3532 3533 3534 3535 3536 3537	1	Gray Iron Foundries
		2	General Industrial Equipment
		3	Plastics
53 Machine Tools and Dies	3541 3542 3544 3545 3546 3547 3549	1	Gray Iron Foundries
		2	Aluminum Foundries
		3	Plastics
54 Special Industrial Equipment	3551 3552 3553 3554 3555 3559	1	Gray Iron Foundries
		2	General Industrial Equipment
		3	Plastics
55 General Industrial Equipment	3561 3562 3563 3564 3565 3566 3567 3568 3569	1	Gray Iron Foundries
		2	Radio, TV, Telephones
		3	Plastics
56 Service Machines	3572 3573 3574 3576 3579 3581 3582 3584 3586 3589 3633	1	Electronic Components
		2	Radio, TV, Telephones
		3	Gray Iron Foundries
57 Radio, TV, Telephones	3612 3613 3621 3622 3629 3631 3651 3661 3662 3671 3672 3673 3693 3694 3699 3825 3822	1	Electronic Components
		2	Gray Iron Foundries
		3	Copper Wire
58 Household Appliances	3623 3624 3632 3634 3635 3636 3639 3641 3643 3644 3645 3636 3647 3648 3652 3691 3692	1	Plastics
		2	Gray Iron Foundries
		3	Radio, TV, Telephones
59 Electronic Components	3674 3675 3676 3677 3678 3679	1	Plastics
		2	Metal Bolts and Wire
		3	Radio, TV, Telephones
60 Motor Vehicles	3483 3711 3713 3714 3715 3721 3724 3728 3761 3764 3769	1	Metal Bolts and Wire
		2	Gray Iron Foundries
		3	Plastics

(continued)

Appendix Table 11.5 Industries Included in the NIT and EDD Systems

Industry	SIC Codes	Rank	Linked Industry
61 Ship and Rail Building	3731 3732 3743	1	Gray Iron Foundries
		2	Engines
		3	General Industrial Equipment
62 Measuring Instruments	3811 3822 3823 3824 3829 3841 3842 3843 3851 3861 3873	1	Plastics
		2	Electronic Components
		3	Gray Iron Foundries
63 Jewelry and Musical Investments	3911 3914 3931	1	Zinc Smelting
		2	Burial Caskets
		3	Nonferrous Smelting
64 Burial Caskets	3915 3995	1	Gray Iron Foundries
		2	Nonferrous Smelting
		3	Cutlery and Hand Tools
65 Sporting Goods	3942 3944 3949 3951 3952 3953 3955	1	Plastics
		2	Gray Iron Foundries
		3	Paper Containers
66 Advertising Signs	3961 3962 3963 3964 3981 3982 3991 3993 3996	1	Plastics
		2	Gray Iron Foundries
		3	Aluminum Foundries
67 Miscellaneous Manufacturing	3983 3999	1	Paper Containers
		2	Plastics
		3	Miscellaneous Metal Products
68 Apparel	2311 2312 2313 2314 2315 2316 2317 2318 2321 2322 2323 2327 2328 2329 2331 2335 2337 2339 2341 2342 2351 2352 2361 2363 2369 2371 2381 2384 2385 2386 2387 2389	1	Broadwoven Fabric
		2	Narrow Fabric
		3	Advertising Signs
69 Military Tanks	3795	1	Gray Iron Foundries
		2	Aluminum Foundries
		3	Copper Wire

Choice Indices for the Apparel Industry

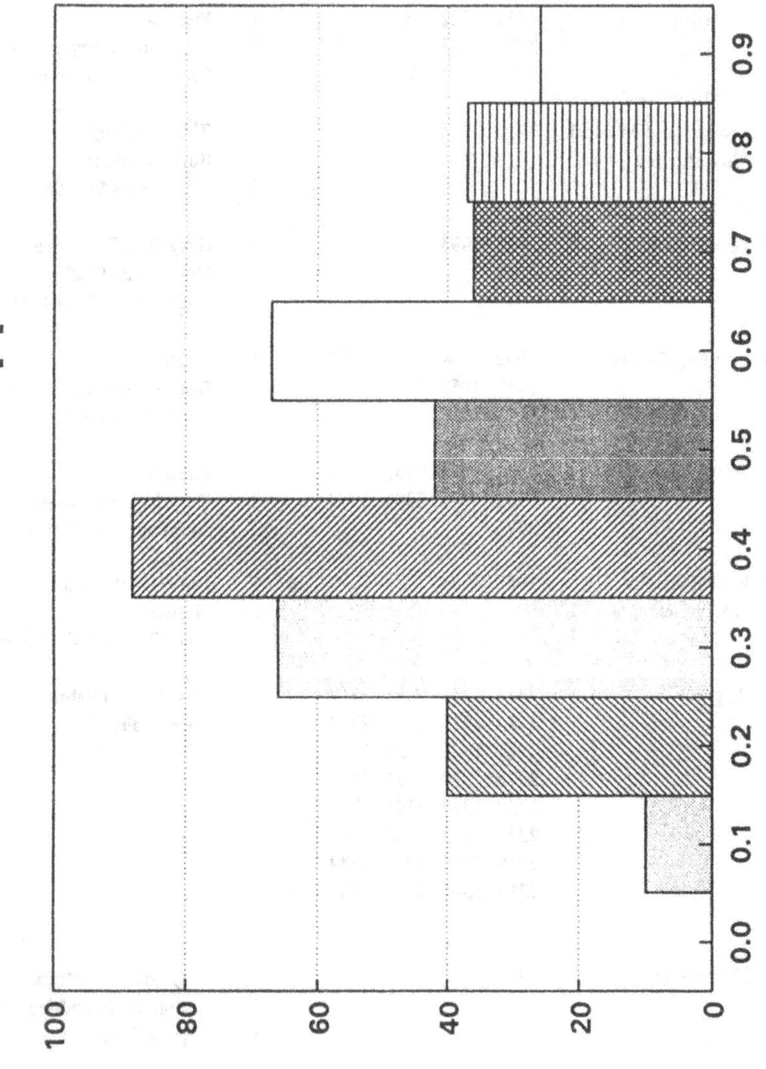

Figure 1

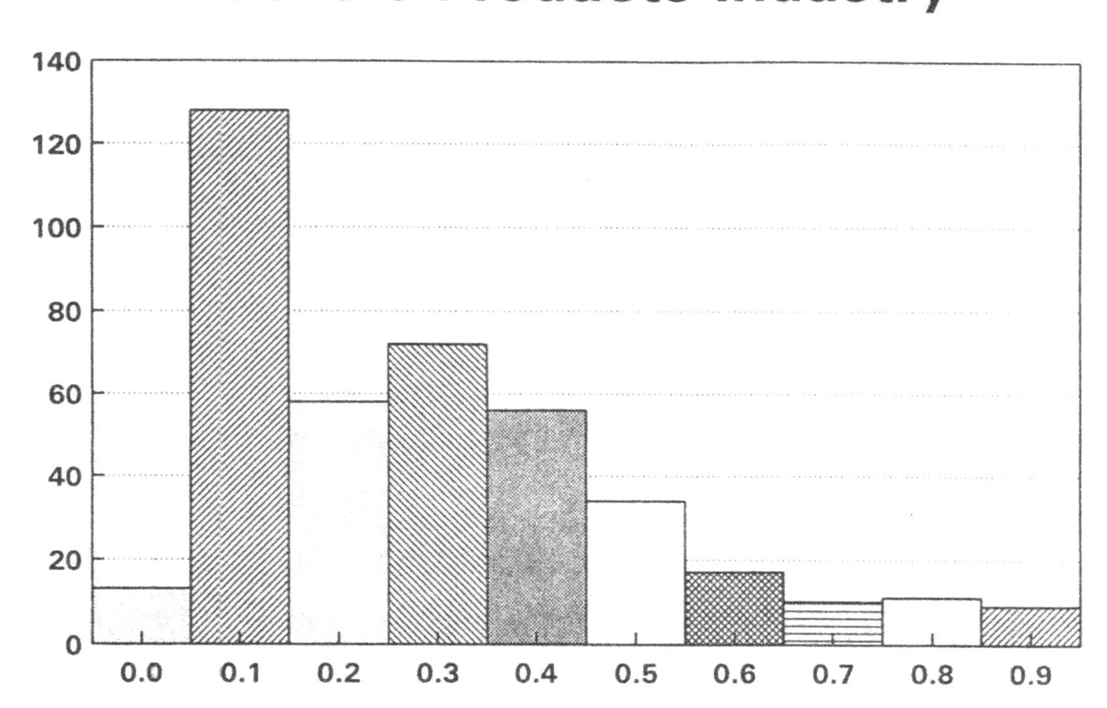
Choices Indices for the Fabricated Textile Products Industry
140
120
100
80
60
40
20
0
0.0
0.1
0.2
0.3
0.4
0.5
0.6
0.7
0.8
0.9

Figure 2

Frequency Distribution of Per Capita Local and County Taxes

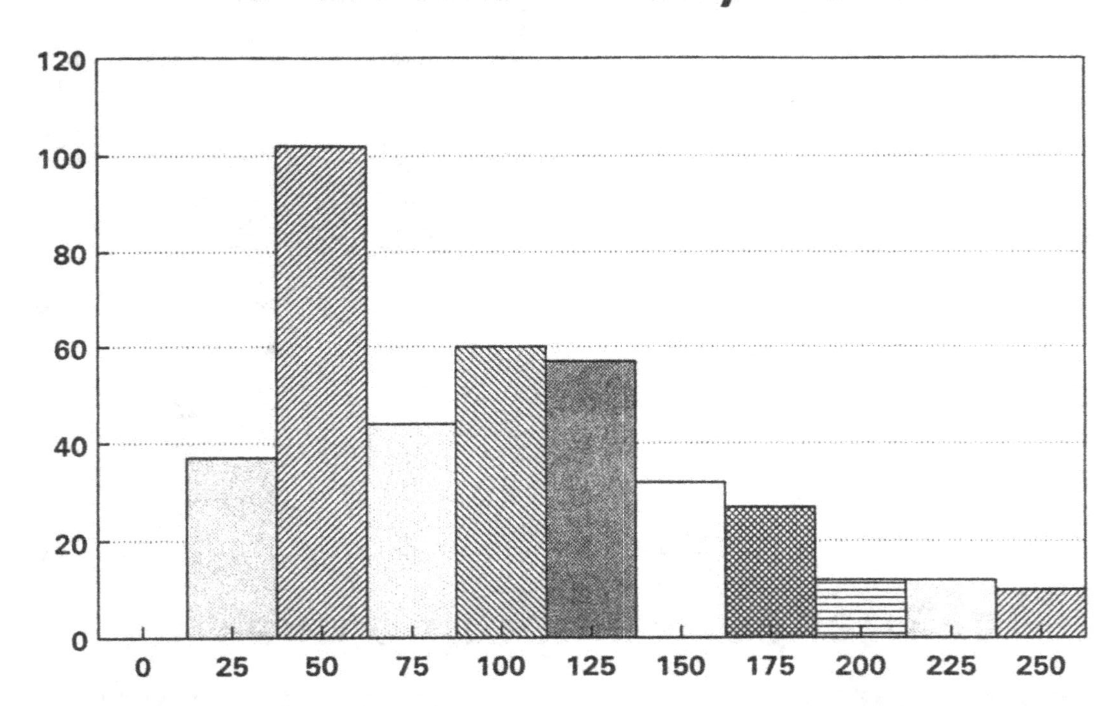

Figure 3

12

Using Input-Output for Regional Planning

Gerald A. Doeksen and Mike D. Woods

Introduction

The changing nature of rural economies has offered new challenges and opportunities for extension and applied economists. The energy boom in the early 80's created growth planning problems. Conversely, the energy and agricultural crisis caused even more severe planning problems for community leaders in Oklahoma. To assist rural leaders, a simulation model developed around an input-output framework has been created. The model was designed not only to measure changes in the economic base but to assist in community service planning. The overall objective of this paper is to provide an overview of the model and to present an application of the model. More specifically, the paper:

1. provides an overview of the community simulation model;
2. presents an application of the model;
3. summarizes the types of problems the model has been used on; and
4. discusses future directions of model refinement.

The Community Simulation Model

The simulation model described here was originally developed in Oklahoma (Woods and Doeksen). The model has been used by community development professionals in both Oklahoma and Texas.

The data necessary for the model is contained in several accounts which are linked through a series of equations. Figure 1 presents an overview of the social accounts contained in the simulation model. The model contains five accounts: an economic account, a capital account, a demographic account, and accounts for community services and community revenue.

The driving force is the economic portion of the model which includes a community specific input-output model and a gravity model. The gravity model is employed to determine the service area of a community based on population levels and distances to nearby communities. A location quotient technique is applied to a regional or state input-output model to derive a community specific input-output model. The community model is made dynamic through the use of equations which predict final demand over time.

The capital account allows for the simulation of investment and its effects on the economy. Capital transactions by industry sectors are included in the capital coefficient matrix. Capacity levels and capital-output ratios describe the relation between capital investment and industry output. This capital data is related to the interindustry information included in the input-output model.

The demographic account contains information on community population. A growth model (based on distance to other communities and population) is utilized to estimate the community service area. A cohort survival submodel predicts population by age-sex categories based on birth rates, death rates, and migration. Population information is stored for both the community and the service area.

The community service account contains usage coefficients for services provided in the community. Services analyzed include hospitals, clinics, emergency medical service, fire protection, water, sewer, and solid waste. Community requirements for each of these services are estimated based on model output. The community service information is based on research conducted for each service in Oklahoma.[1] The revenue account provides projections of local revenue by source such as sales tax, licenses, permits, and user charges for various services. The revenue projections are based on community specific revenue data available for Oklahoma communities.

Figure 2 presents an overview of major computations contained in the model. The economic account contains a local input-output model. Equations for each category (households, capital investment, inventory change, federal government, state and local government, and exports) predict final demand over time. Production relations then determine output levels by economic sector. Labor productivity rates are used to estimate employment requirements by sector. At the same time, the demographic account is estimating population using an age-sex cohort survival technique. Using local labor force productivity rates, the available labor force is then estimated for each year. Net migration is then the balancing variable to match employment needs with labor force. The resulting final population values are then included in the next year's calculations.

The complete model has over 200 equations describing the economic and social relations within a community.[2]

The model is designed to be easily adapted to a wide range of community applications. Specific information on the community is requested for population variables, employment data, and geographic location. This data is readily available from census publications and state employment agency reports. A large secondary data base is included with the model to minimize data collection. Growth rates, input-output parameters and community service coefficients are included in this data base. The computer model is interactive and asks a series of questions to which the user responds by providing the input data required. The model is written in FORTRAN and compiled on an IBM 370/168 computer. When the model was originally developed, the data base and equations required so much storage space that a main frame computer was required. However, given the rapid developments in the microcomputer field, conversion to a microcomputer may be possible in the future. Table 12.1 provides a summary of the information provided by the model.

An Application of the Community Simulation Model

The model provides annual projections of economic and demographic variables for a community. Projections are provided for a baseline run and impact run. The baseline run is based on past growth trends, and projects variables assuming there will be no outside shocks or changes in the local economy. The impact run incorporates information from a major change in the community such as a new industrial employer may move to the community.

Table 12.2 through Table 12.6 provide selected results from an application of the model. The results are for Miami, Oklahoma, a small community with a 1980 population of 15,084. Community leaders, thru a special committee appointed by the city council, requested that OSU conduct a detailed community simulation study. The request was motivated by the Goodrich Tire Company announcing that it was closing down a plant which employed 1,900 workers. The city council wanted the best possible information to plan for the drastic cut in its economy. The study which was completed in December 1985 is presented here to illustrate the model and to evaluate the model predictions.

The model provides annual projections of economic and demographic variables for Miami and its service area. The gravity model determined the service area. The competing communities of Joplin, Vinita, Neosha, and Columbus were first identified. The service area is subsequently slightly smaller than the county. First, a baseline simulation model is generated by running the model as if no changes are expected to occur in the service area from 1970 to 1990. The model results from 1970 to 1985 were compared with Miami and

Ottawa County data to insure that the model was accurately reflecting the Miami service area. The simulation results from 1985 to 1990 assume that the Goodrich plant continues in operation and no other large economic base change occurs.

Then, three additional scenarios were developed by entering certain assumptions about the closing of the Goodrich plan into the simulation model. Each scenario assumes a different number of employees moving from the area. The selected scenarios are based on information obtained from the survey completed by Goodrich personnel.[3]

The Four Scenarios

Survey results are used to construct the four scenarios. These are summarized in Table 12.2. Scenario 1 is the baseline run and assumes the plant does not leave. Scenario 2 assumes only the survey respondents who said they would move from the Miami area will migrate. Scenario 3 assumes that the survey respondents who said they would move as well as those who responded not sure, will migrate from the area. Finally, scenario 4 assumes that the respondents who said they would move and half of those who responded not sure, will migrate from the area.

Employment, Income and Population

Employment projections for the four scenarios are presented on Figure 3. Total wage and salary, and proprietor employment is projected to go from 13,143 in 1980 to 15,547 in 1990. This projects an increase of about 1,400 jobs from 1985 to 1990. The largest share of the growth is projected to be in the service sectors. Employment for scenario 2 is projected to go from 14,183 in 1985 to 11,768 in 1990. This scenario assumes the plant will close and only those surveyed who said they would move will migrate from the area. Scenario 3 projects decreases of employment from 14,183 to 9,374. Scenario 4 depicts those who said they would move and half of the not sure respondents will migrate from the area. Under these assumptions, employment is projected to go from 14,183 in 1985 to 10,502 in 1990.

The 1985 unemployment rate was 10.3 percent. Based on Scenario 4, unemployment was expected to jump to 21.8 percent in 1986, increase further in 1987, and start to decline slightly in 1988, 1989, and 1990 (Table 12.3).

Projected proprietor income, wage and salary payments are on Figure 4. The data reflect the same pattern as displayed by the employment data.

Projected population estimates for the service area and the commmunity are provided in (Doeksen et al.). Without the plant closing, the Miami and service area population is projected to go from 32,465 in 1985 to 32,557 in 1990. Under Scenario 2 the 1990 population is projected at 29,867, and 22,683 under

Scenario 3. For scenario 4, the population of the Miami service area is projected at 26,049 in 1990. The population estimates are provided by 12 age groups and by males and females.

Sales Taxes and Utility Use

Data in Table 12.4 reflects local sales tax collection information from 1980 through 1985. From 1980 through 1984, the sales tax rate was 2 percent and the data reflects actual collections. The 1985 figure reflects sales tax collections for the first nine months of 1985. The remaining projections derived from the model assume a 3 percent collection rate. Without the plant closing, sales tax collections are estimated at $2,512,161 in 1986 and $3,640,815 in 1990. Based on Scenario 4, sales tax collections are projected to be $1,803,366 in 1986 and $2,074,698 in 1990. At this rate, sales tax collections in 1990 will exceed actual collections in 1984.

Projected decreases in water, electrical, sewer and solid waste use are shown in Table 12.5. Usage is projected to decrease from the 1985 level by 2.3 percent in 1986 and by 19.8 percent in 1990. These projections are based on Scenario 4 and reflect the anticipated decrease in residential population and small businesses. These impacts may be overstated if some customers are large industrial users.

Housing

Housing will be a concern of many Miami residents. Data was gathered from the city of Miami concerning housing starts and demolitions. Starting with the 1980 number of housing units and adding new starts and subtracting 30 demolitions per year, an estimate of 6,125 houses in 1985 was made. From 1986 to 1990, it is assumed that 10 new start and 20 demolitions occur. Based on Scenario 4, the 1985 vacancy rate of 4.8 percent is projected to exceed 20 percent in 1990 (Table 12.6).

School Projections

A decline in population will impact all services. School enrollment was singled out to illustrate the adjustments these services will have to make.

The impact on school enrollment is projected in Table 12.7. Since those migrating often have children, the impact on the school enrollment will be greater than on hospital usage. It is projected that school enrollment will be 97 percent of the 1985 enrollment in 1986 and 78 percent of the 1985 enrollment in 1990.

Other Model Results

Decision makers after seeing and reviewing the above results requested two additional scenarios. Scenario 5 for the Miami application is based on the assumption that 100 new jobs would be created by new manufacturing from 1986 through 1990. Scenario 6 assumed that 200 jobs would be created each year from 1986 through 1990. These results were provided in the final report to the community.

It is difficult to compare a specific scenario with what actually happened. However, during the first two years, the projected employment, income and unemployment rates were very close. Since then, the city has been able to attract several industries which are not comparable with scenarios 5 and 6.

Summary of Other Applications

The agricultural and petroleum crisis in Oklahoma has helped generate a large volume of requests from extension clientele. The model is useful for growth and decline issues. The most frequent request is for impact studies related to manufacturing plants opening or closing. Other requests included impact studies for:

1. a proposed national park (Doeksen et.al., 1986);
2. a community losing a medical doctor (Doeksen et.al., 1987);
3. a community losing a hospital (Doeksen and Loewen);
4. a community losing a livestock sales facility (Doeksen et.al., 1987);
5. the results of a county seat town losing 20% of its farmers (Doeksen, 1987); and
6. the impact of a golf course (Doeksen, 1988).

Future Model Plans

The simulation model presented above is extremely complex and takes a very knowledgeable regional economist to apply. In almost all cases, the characteristics of the community and problem are unique such that model changes need to be made. In most cases, it takes a person familiar with the model two weeks to apply the simulation model to a community problem. In addition, the model uses a location quotient technique applied to a national or state I-O table to derive a community I-O model. These limitations have caused the authors to begin developing a more simplified version of the simulation model. The major changes is that IMPLAN will be incorporated into the model such that the model will always contain the latest and most applicable input-output model. The authors are also attempting to simplify the other equations in the model to arrive at a version which can be run on an IBM personnel computer. This will allow the authors and other Extension workers to reach a wider audience in a more timely manner.

Summary

Extension professionals and other professionals working in the area of rural development must have the best economic and planning tools to assist rural leaders as they plan for economic growth or decline. This paper has presented a community simulation model and application of a manufacturing plant loss to a community.

The computerized model is a useful aid for community leaders and supplements other decision aids. For example, the model can aid in planning capital expenditures in cases of growth. It can be used to estimate the need for a specific service such as water or schools. The model enables community leaders to conduct impact analysis as well. The model's flexibility allows adaption to different economic and social circumstances.

Notes

[1] For a detailed description of community service research conducted for Oklahoma communities, see Doeksen and Nelson.

[2] For a complete description of the model and equation used, see Woods and Doeksen (1983) and Woods, Doeksen, and Nelson.

[3] The survey and results are discussed and presented in Doeksen et.al.

References

Doeksen, Gerald A. et al. *"The Economic Impact of the Closing of the Goodrich Plant in Miami Oklahoma."* Department of Agricultural Economics, Paper 85153, Oklahoma State University, Stillwater, Oklahoma, December 1985.

Doeksen, Gerald A. et al. *"An Analysis of the Economic Impacts of the Proposed Prairie Park Preserve on the Pawhuska Area."* Department of Agricultural Economics, Paper 8611. Oklahoma State University, Stillwater, Oklahoma, February 1986.

Doeksen, Gerald A. et al. "Rural Physicians Make Good Economic Sense." Paper presented at National Rural Health Care Association in Nashville, Tennessee on May 6-9, 1987.

Doeksen, Gerald A. and Ron Loewen. *"A Rural Hospitals Impact on a Community's Economic Health."* Department of Agricultural Economics, Professional Paper 2739, Oklahoma State University, Stillwater, Oklahoma, March 1988.

Doeksen, Gerald A. et al. *"The Economic Impact of the Livestock Facility on Hugo, Oklahoma."* Department of Agricultural Economics, paper 8755, Oklahoma State University, Stillwater, Oklahoma, June, 1987.

Doeksen, Gerald A. "The Agricultural Crises as it Affects Rural Communities." *Journal of The Community Development Society.* Vol. 18, No. 1, 1987, pp. 78-88.

Doeksen, Gerald A. *"The Economic Impact of a Golf Course on the Economy of McAlester, Oklahoma."* Department of Agricultural Economics paper 8871, Oklahoma State University, Stillwater, Oklahoma, 1988.

Doeksen, Gerald A. and J.R. Nelson, "Community Service Budgeting: An Effective Extension Tool," Department of Agricultural Economics, Paper 8156, Oklahoma State University, May 1981.

Woods, Mike and Gerald A. Doeksen. *"A Simulation Model for Rural Communities in Oklahoma."* Oklahoma State University, Stillwater, Oklahoma, Agricultural Experiment Station, Bulletin B-770, October, 1983.

Woods, Mike D., Gerald A. Doeksen, and James R. Nelson, "Community Economics: A Simulation Model for Rural Development Planners," *Southern Journal of Agricultural Economics*, Volume 15, Number 2, December 1983, pp. 71-77.

Table 12.1. Summary of Annual Projections Provided by the Community Simulation Model

Category	Model Output
Economic	Employment by Industry Sector Income by Industry Sector Output by Industry Sector
Demographic	Population by Age-Sex Cohort Population for Community and Service Area
Service	Hospital Bed Days by Age-Sex Cohort Physician Visits by Age-Sex Cohort Ambulance Calls Number of Fires Water Requirements Sewer Volume Solid Waste Volume
Revenue	Community Revenue by Source

Source: Oklahoma State Community Simulation Model.

Table 12.2. General Information About Community Simulation Runs

Scenario 1	Baseline run with no changes in economic base of Miami area
Scenario 2	Goodrich plant closing and only survey respondents who said they would move will migrate out of area [1]
Scenario 3	Goodrich plant closing and survey respondents who said they would move and those who were not sure will migrate out of area [2]
Scenario 4	Goodrich plant closing and survey respondents who said they would move and half of not sure respondents will migrate out of area [3]

Source: Oklahoma State Community Simulation Model

[1] (71 + 122)(193 ÷ 1,122 = 17.2%) .172 x 1,900 = 326 employees

[2] (71 + 122 + 205 + 560 = 958)(958 ÷ 1,122 = 85.3%) .853 x 1,900 = 1,620 employees

[3] (71 + 122 + 382 = 575)(575 ÷ 1,122 = 51.2%) .512 x 1,900 = 972 employees

Table 12.3. Projected Unemployment Rate for Scenarios 1 and 4

Year	Scenario 1	Scenario 4
1985	10.3	10.3
1986	10.5	21.8
1987	10.6	24.5
1988	10.7	24.0
1989	10.8	23.6
1990	10.7	23.1

Source: Oklahoma State Community Simulation Model.

Table 12.4. Projected Sales Tax Collections Under Scenarios 1, 2, 3, and 4

Year	Scenario 1	Scenario 2	Scenario 3	Scenario 4
1980	$1,568,867*			
1981	1,687,743*			
1982	1,715,509*			
1983	1,968,928*			
1984	2,023,978*			
1985	1,140,489**			
1986	2,512,161	$1,803,366	$1,803,366	$1,803,366
1987***	2,753,766	1,843,848	1,772,154	1,809,630
11988***	3,020,328	1,983,186	1,779,507	1,883,178
1989***	3,314,691	2,139,219	1,799,064	1,966,824
1990***	3,640,815	2,341,935	1,839,546	2,074,698

Source: Oklahoma State Community Simulation Model.

* Actual Collections

** First Nine Months of 1985

*** Projected from Simulation Model

Table 12.5. Impact on Water, Electrical, Sewer and Solid Waste Revenue[1] (Scenario 4)

Year	Percent Reduction (From 1985)
1986	2.3
1987	7.8
1988	12.5
1989	16.6
1990	19.8

Source: Oklahoma State Community Simulation Model.

1 Revenue Collected for Water and Electrical Consumption is Provided in Appendix Table.

Table 12.6. Projected Vacancy Rate for Miami (Based on Scenario 4)

Year	Housing Units	Vacancies	Percent Vacant
1980[1/]	5,943	302	5.1
1985[2/]	6,125	293	4.8
1986[3/]	6,115	412	6.7
1987[3/]	6,105	728	11.9
1988[3/]	6,095	991	16.2
1989[3/]	6,085	1,216	19.9
1990[3/]	6,075	1,396	22.9

Source: Oklahoma State Community Simulation Model.

1 From Census of Housing.
2 Starts with 1980 Census Figures, adds in new starts from Table 21 and subtracts a demolition estimate of 30 per year.
3 Assumes 10 new starts and 20 demolitions each year.

Table 12.7. Projected School Enrollment for 1986-1990 as Percent of 1985 Enrollment

Year	Percent
1985	100
1986	97
1987	91
1988	86
1989	82
1990	78

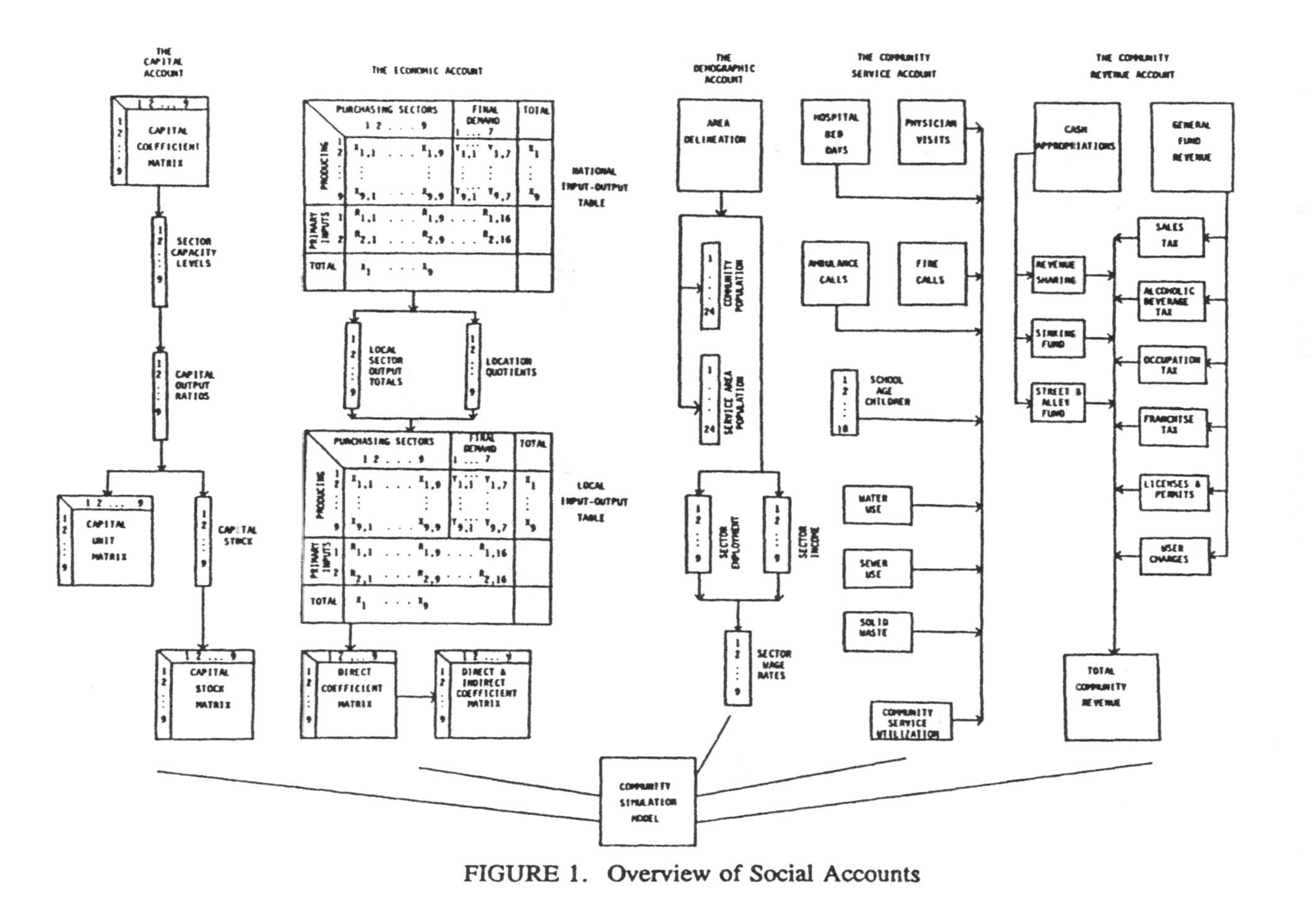

FIGURE 1. Overview of Social Accounts

Figure 2. Overview of Major Computations in the Community Simulation Model

Community Simulation Model

FINAL DEMAND
PREDICTIONS

INITIAL
POPULATION

OUTPUT BY
SECTOR

AGE-SEX COHORT
SURVIVAL

EMPLOYMENT
BY SECTOR

AVAILABLE
LABOR FORCE

IN-MIGRATION AND POPULATION LEVEL

COMMUNITY SERVICE NEEDS

Figure 3. Projected Employment for Miami Area Based on Four Scenarios.

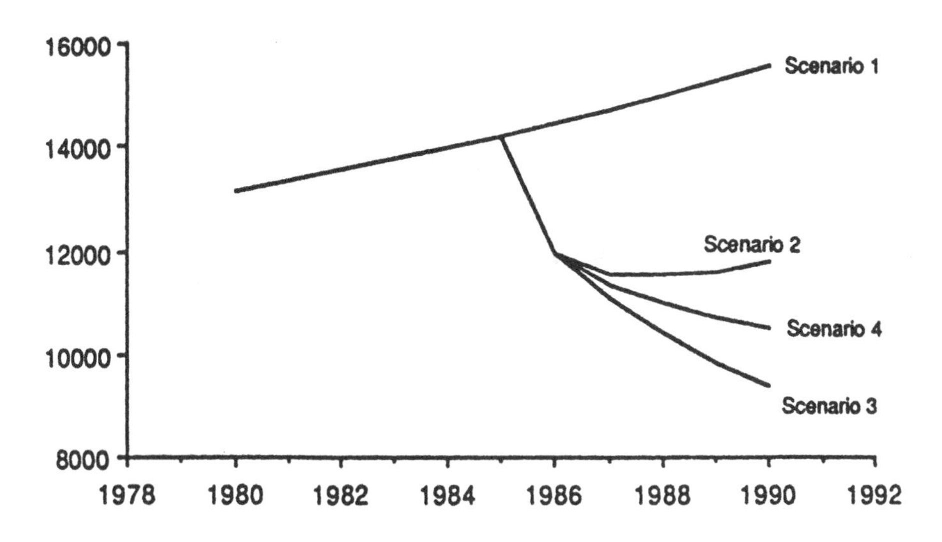

Figure 4. Projected Proprietor Income and Wage and Salary Payments for Miami Area Based on Four Scenarios.

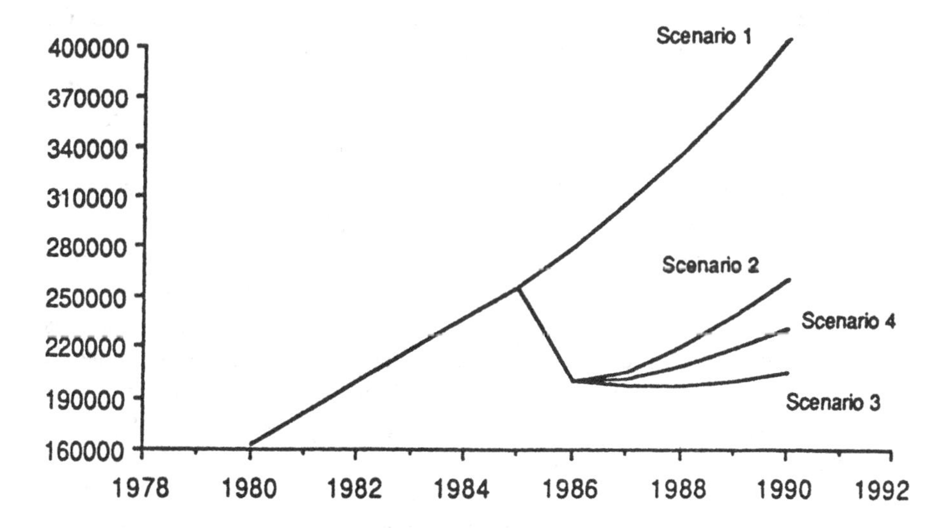

13

Policy Simulation Modeling

George I. Treyz

Introduction

The objective of policy modeling is clearly the accurate assessment of the short and long term effects of an external event on a regional economy. While it is easy to state the objective, the implementation of this objective is not so straightforward. As in the case of all economic modeling, we have to decide where we can simplify without altering the results substantially, and where we must direct our attention in terms of making our assumptions realistic. In the I/O framework, we have found that major attention has to be paid to the proportions of local use supplied locally, i.e., Regional Purchase Coefficients (RPC's). Whereas, in terms of model outcomes the technological coefficients at a disaggregated level can be drawn from other sources, in most cases, without substantially affecting the results. In terms of the larger modeling questions, people using standard input/output models have implicitly or explicitly decided that certain features of neoclassical economics are unimportant and that it is more important to develop the techniques for refining the input/output structure itself. It is here that I would disagree. In my opinion the implicit assumptions behind I/O of an infinite supply of labor at the going wage, for example, is sufficiently unrealistic so that it may completely alter the results we get from a policy simulation. Therefore, I will propose in this paper that in order to do realistic policy assessment, it is necessary for us to go beyond economic base and input/output models. We need to introduce features into these models which will allow some response: of exports to changes in the relative cost of doing business, of wage rates to changes in labor market conditions, of migration to shifts in economic conditions in the area and of the relative use of factors of production to changes in their relative costs in the production process.

In Figure 1, we can see the economic assumptions of the I/O framework in diagrammatic form. Here I have shown a simple case where, for example, there is only one intermediate good industry or where the requirements of intermediate or final export demand are the same, that we can draw a diagram similar to the familiar Keynesian diagram. In this case, we see that the first line above the horizontal axis shows that employment dependent on demand for intermediate inputs (EINT) grows proportionally to the growth of employment in the area. The second line above the horizontal axis shows that if we add to intermediate employment, the employment induced by respending of income (EIDU), this sum is proportional to total employment in the area. To the extent that income is proportional to employment and consumption derives directly from income, this is a reasonable representation of the induced plus intermediate demand in economic base or I/O models. Finally, we add export employment (EXPT) and, assuming this export-dependent employment is exogenous to the system, we draw this as a line parallel to the line which sums up induced intermediate employment. By drawing a 45-degree line from the origin, we can show that the system will be at equilibrium when total demand for intermediate, induced and export employment is equal to total employment. This occurs where the 45-degree line meets the total demand for the three kinds of employment. This model can therefore be used to determine total employment at equilibrium as shown in Figure 1. Below the 45-degree line in the diagram in Figure 1, I have drawn another diagram indicating that in the input/output and economic base models we assume that the wage rates are fixed and that employees are available at the going wage rate as required. In this diagram I have also shown that an increase in export employment will lead to a multiple increase in total employment. This is shown in the wage and total employment graph beneath the 45-degree line diagram, as well as in the first figure in the diagram.

In Figure 2 I have constructed the same 45-degree line diagram, but in this case I have shown a diagram directly below it which is more consistent with traditional neoclassical assumptions about the workings of a regional economy. Again, as in Figure 1, an exogenous increase in the economic base shifts up the total demand line, but in this case instead of shifting total employment from ET_0 to ET_1 it shifts the demand curve for employment at wage rate W_0 by the distance between ET_0 and ET_1. From diagram 2 the figure under the Keynesian diagram clearly indicates that the equilibrium point will not be at ET_1, but rather will be at ET_2 which is clearly less than ET_1. Neoclassical economists would point out that there are at least two reasons why the demand curve for employment may be expected to be downward sloping to the right rather than vertical as is implied in the economic base and input/output framework. The first reason is that as wage rates increase, capital and other factor inputs are substituted for labor. The second reason is that in a regional economy as costs increase, producers selling in the national marketplace will be less competitive. Either they will raise their prices and lose some sales or some manufacturers may find it profitable to move their production to another location when costs go up in the area in question. These two reasons then argue for a downward sloping demand curve as shown in Figure 2 rather than the vertical one shown in Figure 1. An additional reason might include import substitution when relative prices increase (i.e., a reduction in the RPC's).

Economists could also argue for the upward sloping supply curve as shown in Figure 2 for a regional economy on two bases. One is that as the wage rate increases, one may find some substitution of work for leisure. While this is not a theoretical certainty, given the possibility of the income effect overcoming the substitution of leisure for labor effect, it is a possible result. The more important reason for the upward sloping supply curve is surely that if the participation rate is not raised significantly by increased wages, then it will be necessary to attract new migrants into the area. Since migrants move for economic as well as other motives, it is reasonable to believe everything else being equal, an increase in the relative wage in the area is necessary to attract more migrants than otherwise would have come to this particular region. In the short run, the elasticity of supply will be much lower than in the long run. However, even in the long run given higher land and housing prices as the population increases, we could expect that an increase in the nominal wage would be necessary to attract a larger labor force. Thus, the upward slope for the supply curve, indicating that an increase in wage may be necessary to change the participation rate and to change the size of the population, seems reasonable. Even in areas with substantial unemployment, a reduction in the unemployment rate below previous levels is often associated with a tighter labor market and increases in the real relative wage in that area over and above what it would be if the high unemployment conditions had continued to prevail. It is my opinion, based in part on our quantitative work, that in many cases ET_1 may be a misleading approximation of ET_2. In fact, I would go further to assert that the likely difference between ET_2 and ET_1 in the short to intermediate run is quantitatively likely to be more important than the gain in accuracy we are likely to get from taking extraordinary measures to further refine the technological coefficients or even our techniques of regionalization in the I/O framework.

Thus, the approach of policy simulations modeling is to interface neoclassical and other features into the present highly developed I/O framework. The purpose is to provide a more accurate basis for policy simulations. By building on the I/O structure, it is our intention to incorporate all of the strengths of the I/O analysis while mitigating some of the inaccuracies that may occur if we rely too heavily on the standard assumptions made in the I/O model.

In constructing an I/O based policy simulation model with neoclassical responses, we can build a complete model of the economy that one can use historically to construct variables, such as the relative cost of doing business by industry, that are unavailable in consistent, uniform fashion to regional modelers. Having constructed a history, we can then generate a control forecast. After the forecast is in place, we can introduce a policy change into the system by changing any policy variable. This then will make it possible to generate a complete alternative forecast. Then, by subtracting the control forecast from the disturbed alternative forecast, we will be able to show the effects of the policy action on the path that the regional economy is likely to follow. In this paper I will discuss the strengths and weaknesses of the I/O framework for models to be used in economic development analysis, and then I will discuss how one can build onto the I/O framework.

Strengths and Weaknesses of I/O Models for Economic Development

The I/O structure provides a good framework for regionalization. By starting at a very disaggregated level and calculating supplies and demands for each industry, it is possible to estimate I/O coefficients assuming no crosshauling. Since crosshauling is a major fact in the operation of any regional economy in the U.S., it is of course necessary to go beyond this approach and to use an approach for regionalization that will take into account the clear and substantial crosshauling that one finds when one looks at the Census of Transportation. By using an equation based on these observed data and variables in addition to the supply demand ratio available in the regional data, it is possible to make a good estimate of the RPC's that are required to properly regionalize a model. With these coefficients at the very disaggregated level, aggregation can proceed from this point for those wishing to work with a somewhat more aggregated model.

In addition to providing a framework for calibrating a model initially to a region, the I/O structure is very important in showing how changes in the export base can influence the economy. By identifying the export base in great detail, one can see what makes the local economy tick and what are possible avenues for increasing or decreasing the size of the regional economy as desired by policymakers and people in the area. In addition to its usual application showing how changes in exports from particular industries can influence the economy, the I/O framework can also be used to indicate how import substitution would increase economic activity in the area.

Of course the I/O model is much superior to the economic base model in that we can observe different multipliers for different industries. These multipliers can be very different and, therefore, the effort of dividing the economy into industrial sectors proves to be very worthwhile if one is to make an accurate estimate of the effects of a policy on the area. In addition to providing more accurate multipliers through disaggregation, the detail of the simulation itself can be of great importance to policymakers, whether this detail helps to show what industrial facilities will be needed for supply industries or whether it is used to help indicate the occupational demands that will be prevalent in the area. Offsetting this strength in the I/O models, in my opinion, is the possibility of wildly inaccurate results as indicated in my diagrams. In places of tight labor supply, it is not at all impossible that the accurate prediction of the effect of introducing some new economic base activity in the area is a multiplier less than one. This is an easy possibility if the new export industry squeezes out other industries in the economic base or reduces the RPC's of industries in the area due to bidding away scarce labor. The fact is that any multiplier above zero is a possibility and multipliers below one cannot be ruled out on theoretical grounds.

Building on the I/O Framework

It is surprisingly easy to build on the I/O framework if one uses a Cobb-Douglas production function with constant returns to scale. It is a straightforward matter to derive the appropriate cost functions and labor demand functions, as well as capital demand functions, that flow out of the I/O framework. Another advantage of using this production function is that the technological coefficients shown in the I/O framework also can be used as the shares for cost equations for each industry. The fact that in the long run relative costs really do make a difference in factor proportions must be obvious to us all both from observing how relative factor proportions differ in parts of the country or parts of the world where relative factor costs differ, and also from observing how fuel use per unit of output was influenced by a change in fuel prices within the last decade and a half in the U.S.

The next task in introducing a downward sloping demand curve to the I/O analysis is to estimate an elasticity of response in the cost of doing business for exports. In principle, one could estimate this elasticity for each industry in each area. It would be necessary to construct the export series as a residual (by looking at total output in the industry and subtracting from it intermediate demand and induced demand), and then express these exports as a share of U.S. demand for interregional trade in this industry. The resulting share could then be regressed on changes in the relative costs of doing business as they were built up and any other measurable factors that might influence the location of industry. In fact, our time series are rather short and there are many other factors, some intangible and some difficult to measure, that do influence the location of any particular business in any particular state, in any particular year. The combination of noisy series as well as important random factors make it very difficult to extract reliable coefficients from one industry over the time series data that we have available in one region. To overcome this, one can use pooling techniques to estimate the elasticity of export response for individual industries or groups of industries over the larger data set. In the same way that data sets for all states are used to estimate the RPC's, data from all states can be used to estimate elasticities of location response to changes of the cost of doing business in local areas. In this case, we use time series for all industries for all states. Other explanatory variables that are important for two-digit industries include what is happening nationally at a three- or four-digit level to the industries that compose this two-digit industry in the area. Even including this variable, it is necessary to perform individual regression on unexplained trends in particular industries, in particular states.

In trying to estimate the supply curve, the most important factor is the wage response to labor market conditions. Instead of directly estimating a supply curve, our approach examines how wage rates respond to differential labor market conditions. It has also proved useful to split industrial employment down into occupational employment and then look at the response of occupational wages, using a micro data set, to different employment and labor market conditions.

Finally, to fully estimate the upward sloping supply curve it is necessary to explicitly consider how changes in labor market conditions, including employment opportunity, the mix of industries, and real wage rates, may influence the net migration into or out of an area depending on its economic conditions relative to the country.

In short, by building in a flexible production function, export elasticity, wage response, and migration response, it is possible to capture the main features of a model with an upward sloping supply curve and downward sloping demand curve instead of an implicit vertical demand curve and horizontal supply curve of labor.

Summary

In summary, I would like to emphasize the importance of the type of work done at this conference in developing I/O methodology and giving special care to the kinds of applications. I believe that there are very important gains to be made from extending this I/O methodology. However, I would caution that direct application of the standard I/O methodology may be misleading and should be reserved for special situations. I assert this because I believe that some of the simplifying assumptions made in standard I/O work are quantitatively important to the results that one gets when using a model for policy simulation. While emphasizing the need to add other features to the modeling system, I am not an advocate of building more and more elaborate models until we have our feet soundly grounded on the modeling structures that we have been working with. I think that by going too far and relying on econometric estimation of parameters in short time series data for individual regions, we would be building models on an inadequate information base and would produce unreliable simulations. Therefore, the greatest gains may be made by building on the structure that is already in place and by carefully proceeding using regression coefficients which are garnered from pooled analysis across all areas, rather than relying on individual estimates which may suffer from the preponderance of random disturbances in small area economies. In carrying this philosophy of building on the I/O structure model, one could have introduced a way that one can shut off each of these new features in turn until finally one will arrive at a standard I/O framework. Thus, for example, if one has done a policy simulation but then wonders how sensitive the simulation would have been to shutting off the wage response, one can shut off that response and rerun the simulation. By presenting a range of answers based on suppression of different parts of the model, it is possible to show users how the model results depend on various parameter estimates in the model. From these alternative simulations it becomes apparent that the assumptions in standard I/O models represent a very special case of a more general model.

References

Treyz, G.I., A.F. Friedlaender, and B.H. Stevens "The Employment Sector of a Regional Economic Policy Simulation Model," *The Review of Economics and Statistics*, Vol. LXIII, No. 1., February 1980, pp. 63-73.

Treyz, G.I. and B.H. Stevens. "The TFS Regional Modeling Methodology," - *Regional Studies*, Vol. 19.6, 1985, pp. 547-562.

Treyz, G.I., M.J. Greenwood, G.L. Hunt, and B.H. Stevens. "A Multi-Regional Economic-Demographic Model for Regions in the United States," in *Recent Advances in Regional Economic Modeling*. Edited by F. Harrigan and P.G. McGregor. London: London Papers in Regional Science #19, Pion Limited, 1988, pp. 66-82.

Figure 1. Input/Output Model Assumptions

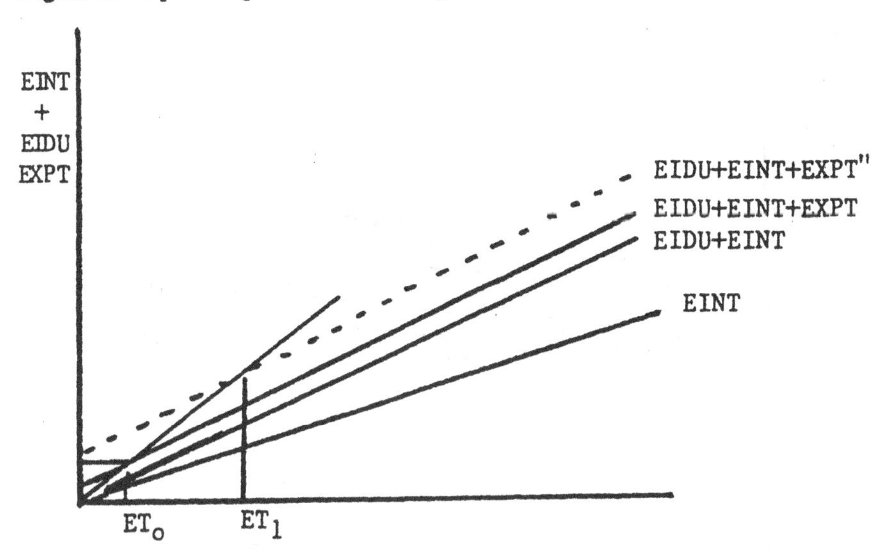

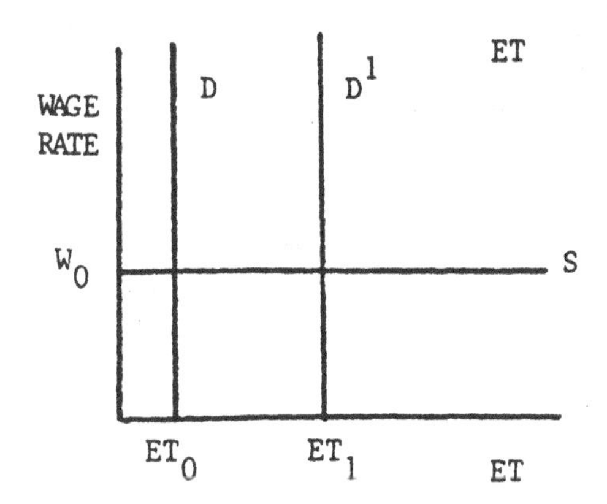

EINT - Employment depending on intermediate input demand
EIDU - Employment dependent on demand induced by local income
EXPT - Employment dependent on exports from the local area
ET - Total local employment
W - Wage rate
---- - Demand including an exogenous increase in exports

Figure 2. Forecasting and Simulation Model Assumptions

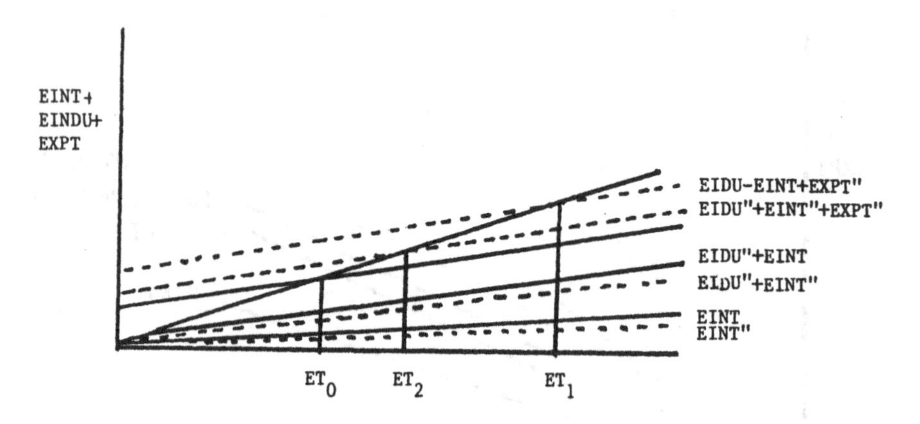

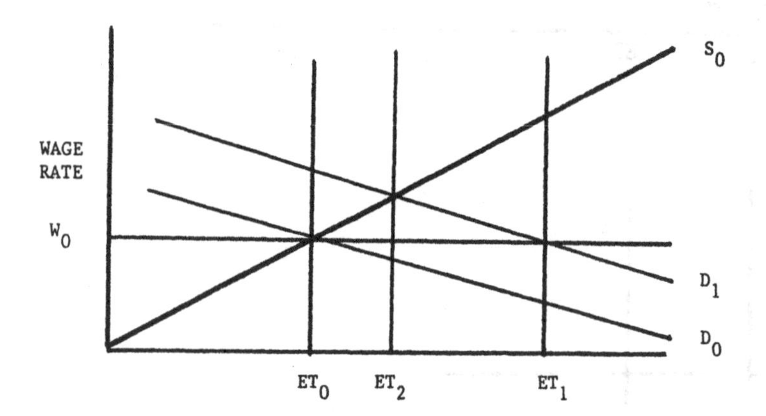

EINT - Employment depending on intermediate input demand
EIDU - Employment dependent on demand induced by local income
EXPT - Employment dependent on exports from the local area
ET - Total local employment
W - Wage rate
---- - Demand including an exogenous increase in exports and the effect of increased wages and costs of production

14

SAM Multipliers: Their Interpretation and Relationship to Input-Output Multipliers

David Holland and Peter Wyeth

Introduction

The focus of this paper is on the social accounting matrix and economic models constructed from social accounting matrices. The approach will be to first review the flow of income as it is captured in a social accounting matrix (SAM) representative of the U.S. economy. From this matrix a simple economic model is constructed, SAM multipliers are calculated, and their economic interpretation is reviewed. There is an extensive literature on SAM models and SAM multipliers (Pyatt and Round, 1985; Robinson and Roland-Holst, 1987), but it can be difficult to relate to input-output models because of its complexity. By working with a very simple and highly aggregated SAM model that is very similar to an input-output model with households endogenous, the relation between SAM and input-output (IO) should be more transparent.

We will conclude with a discussion and comparison between SAM coefficients and (IO) Leontief coefficients when households are treated as endogenous in IO models. This will allow us to illustrate the distinctions between SAM and IO models more clearly as well as to point out the implications of SAM accounts on the issue of model closure in the case of the regional, Type II, IO model constructed from IO accounts.

Structure of the SAM

The SAM Accounts

The basic structure of a SAM follows from the National Income and Product Accounts. Table 14.1 gives a highly aggregated schematic version of the U.S. SAM based on data provided by Economic Engineering Associates (Adelman and Robinson, 1986). The major categories that appear for both the rows and columns of the SAM are production, consumption, accumulation, and trade accounts. These main accounts are broken down into several subaccounts. Although there tends to be considerable variation in the specification of subaccounts for any given SAM, the major accounts in Table 14.1 are common to all SAM's.

Production Accounts. The production accounts are composed of production activities and factors of production. Activities use commodities in the form of goods and services to produce commodities. For the version of the Sam in Table 14.1, separate commodity and activity accounts from a more disaggregated SAM have been combined into activity accounts alone.

The factors of production accounts relate to the primary factors that are used by society in the production process. They are often referred to as the value added accounts which are used extensively in input-output analysis. Traditionally they comprise land, labor, and capital. The factor accounts are paid by activities when production takes place.

Referring to Table 14.1, it is possible to read across the activities row to determine total commodity demand. It is composed of commodities consumed by activities in production, household consumption, government consumption, investment, and exports. The consumption of commodities by activities is referred to as intermediate demand and is used in forming the technical requirements matrix. The activities column show expenditures on inputs used in the production process, value added payments to primary factors, and indirect taxes paid to government. Value added refers to total input purchases of an activity minus its inputs purchased from other activities. Value added consists of payments to households for labor and returns to capital. The sum of all of the inputs used in production must equal gross domestic production at factor cost. The sum of all factor payments comprises gross factor incomes.

These incomes are in turn redistributed to what are called institutional accounts in the value added columns. The rows and columns for factors of production both sum to gross factor incomes and must equal each other so that all the income received by a given factor account is distributed to the institutional accounts.

The institution accounts receive factor income from the value added accounts and distribute it to government, household, or capital (saving) accounts. The enterprises institution represents incorporated business and receives income in the form of returns to capital and depreciation allowances. This institution pays part of these returns back to households in the form of dividends, interest, and rent. Depreciation and retained earnings are the basis for enterprises contribution to the capital or savings row.

Consumption Accounts. The consumption accounts consist of households and government, and are a major component of the final demand accounts.

The columns for the accounts of households, for example, sum to gross expenditures and consist of household expenditures on goods and services, payments of direct taxes, as well as savings and gross transfers abroad. The rows for households represent gross receipts from labor, proprietors income, receipts for capital earnings from enterprises, receipts from government transfers, and earnings from abroad. Gross household receipts must equal gross household expenditures. Household income in the U.S. SAM is distinguished according to the size distribution of income. Often a distinction is made between income going to rural and urban households.

Accumulation Accounts. The accumulation accounts record capital investment and change in stocks in the column and savings from households, enterprises, and government as well as the balance of foreign trade on capital account in the row. The saving from enterprises, households, and government accounts are all combined into one row which shows the source of capital payments. Investment is financed by savings of domestic institutions and foreign financing through the balance of payments, such that gross capital receipts and capital payments equate.

Trade Accounts and the Treatment of Imports. The trade accounts show U.S. economic interactions with the rest of the world. There are two separate trade accounts, one representing outflows of goods and services (exports) and inflows of money; the other representing inflows of goods and services and outflows of money. The trade row shows the outflows of revenue to other countries in the purchase of imports and transfers to abroad from institutions. The trade column shows the inflows of revenue from other countries from the purchase of U.S. exports. Once again, gross payments abroad must equal gross current receipts from abroad.

The commodity rows (1-3) in Table 14.1 show the absorption of goods and services by major components of the economy. The flows represented in these rows are import ridden. In other words, the flows of commodities represent both domestic production and imports. At the bottom of the activity columns, imports by commodity are identified. The column total represents gross total outlay for the activity as well as of imports of the commodity produced by that activity. In this way the sum of each column is equal to the sum of each row for rows 1-3.

As a result of this particular treatment of imports in the U.S. SAM, it is not possible to calculate the technical coefficients necessary to build economic models from the SAM by simply normalizing on the column totals for the activities block. One approach would be to adjust the column total for activities by subtracting commodity imports. The resulting numbers are gross total outlays by activity and the resulting technical coefficients represent the production recipe from both domestic and imported sources for any commodity.

The SAM coefficients in an economic model based on these technical coefficients show the necessary total supply response for a shock in an exogenous variable. By making the assumption that domestic commodities are perfect substitutes for imported commodities the Leontief coefficients could be interpreted as indicating the required change in domestic output.

In order to construct an economic model which shows the relationship between demand shocks and changes in domestic output it is necessary to first purge the commodity rows of their import content. The material balance equation for this SAM may be written as follows:

$$(1) \quad X + M = V + F + E.$$

where:

X = domestic supply
M = imported supply
V = intermediate demand
F = domestic final demand
E = export demand.

In order to construct an economic model of the domestic economy under the assumption that imports are not perfect substitutes for domestically produced goods it is necessary to purge imports from V and F.

A common approach to dealing with the problem is to assume that imports are used in only domestic production and are not directly embodied in exports. The second assumption is that imports are absorbed in both intermediate and final demand (except exports) in a given proportion for every commodity. This proportion may be found by the following equation:

$$(2) \quad I_i = 1 - d_i$$

where:

$$(3) \quad d_i = \frac{X_i - E_i}{X_i + M_i}$$

and:

X_i = domestic output of commodity i,
E_i = exports of domestic commodities i,
M_i = imports of commodity i.
d_i = domestic demand for domestic production of commodity i divided by total commodity supply for commodity i.
X_i = domestic production of commodity i.

A diagonal matrix D of the d_i parameters is then used to purge the material balance equations of imports as follows:

$$(4) \quad Vd = DV$$

and

$$(5) \quad Fd = DF.$$

Then we may write the domestic supply and demand balance equations as:

$$(6) \quad X = Vd + Fd + E.$$

The result of purging imports from the activity rows in the SAM are presented in the SAM depicted in Table 14.2. Imports now appearing in the Rest of the World row are classified by sector of destination rather than by sector of origin as was the case in Table 14.1.

The Construction of SAM Models

To move from a set of accounts to a model requires that additional assumptions be made (Adelman and Robinson, 1986). A common approach in Type I input-output models is to use the fixed coefficients assumption. Under this assumption the elements in each column of the inter-industry accounts are divided by the respective column total resulting in a table of technical coefficients. These coefficients are assumed to represent the production functions of the firms represented by each sector. By assuming that firms respond to changes in demand according to the parameters of the fixed-proportion production function, a model can be specified as a system of simultaneous linear equations. The model can then be solved to yield coefficients through which changes in final demand are translated into changes in each sectors supply (Miller and Blair, 1985).

Similar assumptions are needed when creating a SAM model. Since the SAM model includes a more comprehensive view of the circular flow of income than a standard input-output model, it requires that the fixed coefficients assumption extend to the coefficients of all the endogenous accounts. The fixed coefficients assumption, which in interindustry input-output models is a fixed technology assumption, now must include the assumption that various household expenditure coefficients are fixed when household variables are treated as endogenous.

In input-output accounts only the interindustry linkages are formally specified. The linkage between household income and household spending is not defined nor is the linkage between government revenues and government spending or the linkage between saving and investment. The identification of these linkages in SAM accounts permits industry/household linkages to be specified with the same precision that interindustry linkages are specified in the IO model. The result is that in SAM models, household, government, and investment variables may be more accurately treated as endogenous variables.

For purposes of this paper, only households are treated as endogenous. Our intent is to encourage a connection to a similar type of IO model (Type II) with which many readers will be familiar. In order to construct a SAM model an assumption similar to the fixed coefficients assumption for the input-output model must be made. All of the normalized column coefficients for the endogenous accounts are assumed to be constant in the SAM model. The result is that in addition to the fixed technical coefficients of the IO model, the distribution of nominal income between wages and profits must be assumed fixed, along with the distribution of wage and profit income to households, average tax and savings rates of households, and the sectoral composition of household consumption.

The result of treating households endogenous is a partitioned SAM, (the exogenous columns and rows are excluded) shown as follows (Table 14.1):

$$\begin{array}{l} \text{Activities} \\ \text{Value Added} \\ \text{Endogenous Institutions} \end{array} \qquad S = \begin{bmatrix} A & O & C \\ V & O & O \\ O & Y & H \end{bmatrix} \tag{7}$$

where:

S = matrix of SAM direct coefficients (12,12)
A = matrix of technical coefficients (3,3)
V = matrix of value added (VA) coefficients (3,3)
Y = matrix of VA distribution coefficients (6,3)
C = matrix of expenditure coefficients (3,6)
H = matrix of institutional and household distribution coefficients (6,6).

The supply and demand balance equations can then be written as:

$$(8) \quad \begin{bmatrix} X \\ V \\ Y \end{bmatrix} = S \begin{bmatrix} X \\ V \\ Y \end{bmatrix} + \begin{bmatrix} ex \\ ev \\ ey \end{bmatrix}$$

where:

X = vector of sectoral supply (3,1)
V = vector of value added by categories (3,1)
Y = vector of household incomes (6,1)
ex = vector of exogenous commodity demand (3,1)
ev = vector of exogenous value added (3,1)
ey = vector of exogenous household incomes (6,1)

The (I-S) matrix can then be inverted to specify a matrix equation that expresses levels of sectoral supply, value added, and household income as a function of exogenous variables. This yields:

$$(9) \quad \begin{bmatrix} X \\ V \\ Y \end{bmatrix} = (I-S)^{-1} \begin{bmatrix} ex \\ ev \\ ey \end{bmatrix}$$

where $(I\text{-}S)^{-1}$ represents the matrix of SAM coefficients.

SAM Coefficients

The matrix of SAM direct coefficients is displayed in Table 14.3 while the matrix of SAM inverse coefficients is displayed in Table 14.4. Since we are treating households as endogenous, a similar interpretation of SAM inverse coefficients maintains as would be the case in a Type II IO model. Consider the coefficients in column 1 of Table 14.4. In response to an increase of $1 billion in final demand for agricultural output, an additional output of $.41 billion from agriculture is generated as well $1.97 billion from agricultural related sectors and $.49 billion from other industries.

Payments to labor are increased by $.94 billion, payments to capital by $.73 billion and indirect taxes paid increased by $.15 billion. If we wish to know how households will be impacted, the SAM framework contains the linkage between the payment of wage and profit income and the distribution of that income. It turns out that the income of low income households is estimated to increase by $.14 billion, medium income households will experience an increase of $.50 billion, and high income households will receive an increase of $.58 billion.

An example of the improvement in multiplier interpretation that is offered with SAM models comes from the coefficients in the household columns. Consider an increase in government spending in the form of a transfer to low income households of $1 billion. As a result of this shock, agricultural output will increase by $.11 billion, agriculture related output will increase by $2.63 billion and other industry output will increase by $.64 billion. Payments to labor will increase by $1.08 billion, payments to capital will increase by $.57 billion and indirect taxes will increase by $.16 billion. Perhaps most interesting of all, income to low income households will increase by $.14 billion above the original injection, medium households will experience an increase of $.53 billion and high income households will receive an increase of $.59 billion.

SAM and Input-Output Multipliers: Implications for Regional Models

In regional input-output tables the flow of income to households and other institutions that is detailed in the institutional accounts in a SAM is missing. As a result, regional models treating households as endogenous have a problem with accurate specification of household income (the household row) and the corresponding income base from which consumption and saving decisions are made.

The SAM represented in Table 14.5a takes data from the SAM in the original Table 14.2 and reduces it so that the household accounts are collapsed into one row and column (Pyatt, 1986). The household, government and capital accounts appear as they would in an input-output table if it were possible to trace the payment of value added in each sector to one of the above accounts of final destination. The flow for households includes wages and salaries less social insurance contributions, plus proprietor income and capital income not retained by corporations. Indirect business taxes accrue to government as does a portion of income paid to labor and capital. Households are also shown as receiving income transfers from government and net income from abroad. The resulting sum of the household row is equal to personal income. In the household column, personal income taxes, consumption, and saving are the main flows accounting for the "expenditures" of personal income.

The resulting SAM multipliers treating households endogenous are presented in Table 14.5c. The SAM multipliers for this model provide the target against which other limited information forms of model closure will be compared.

Using Value Added as a Proxy for Income. A commonly used approach to closing regional input-output model with respect to households is to simply use total value added as the estimate of household receipts and as the base on which household saving and consumption decisions are made (Bourque, 1987). The results of closing the model in this way are presented in Table 14.6a. All of the estimates of sector payments to households are considerably overestimated and, as a result, the estimates of household income per unit of sector output are overestimated (Table 14.6b). The household column technical coefficient estimates are underestimated relative to the corresponding SAM estimates in Table 14.5b. This results from the fact that total value added is a much larger figure than personal income, the income base in the SAM. In other words, consumption per unit of income is underestimated in the value added closure.

The Leontief coefficients in the interindustry portion of the model are generally substantially overestimated relative to the same coefficients in the SAM model (Table 14.6c). The range of the overestimate is between 2 percent and 41 percent greater than the corresponding SAM coefficient. This results from using value added as the proxy for payments to households. However, the real damage from the value added approach shows up in the estimates of income multipliers (the household row). Here the overestimate is on the order of 80 to over 100 percent. This results from the overstatement of income to households that stems from using value added as a proxy for household income.

Using Earnings as a Proxy for Income. With this approach, earnings (as defined by the Bureau of Economic Analysis), rather than value added, is used as the proxy for income paid to households. In the household column, personal income rather than total earnings is used as the household income base (Bourque, 1987). The results from this approach to model closure are summarized in Table 14.7a. The household column for the endogenous portion of the model is identical to the household column in the SAM. The household row technical coefficients are underestimated by the earnings estimates, but are of roughly the same order of magnitude as the SAM income payments to households (Table 14.7b). The result is that the earnings model Leontief coefficients are much closer to the true SAM model coefficients than the value added model (Table 14.7c). Except for one income multiplier, all the Leontief coefficients in the earnings model are within ten percent of the SAM model coefficients.

The problem of closure of Type II input-output models has long been recognized. It has been suggested that since the earnings closure underestimates the true income-consumption linkage and the value added model overestimates the income part of that same linkage, that the average of coefficients obtained from the two closure approaches would provide the best estimate of true coefficients.

The evidence presented in this paper doesn't support this argument. The upward bias in the value added closure was much larger than the downward bias in the earnings closure. Averaging the two estimates would result in greater upward bias than just using the results of the earnings approach.

The results of this study suggest that a reasonable approach to the issue of model closure in the case of Type II regional input-output models is to use estimates of sector earnings in the household row. Personal income for the region may be used as the base on which the consumption coefficients in the household column are calculated. Models based on the value added proxy will result in large upward bias in Leontief coefficients and should not be used if the earnings-personal income alternative is available.

References

Adelman, I. and S. Robinson. "U.S. Agriculture in a General Equilibrium Framework: Analysis with a Social Accounting Matrix." *Amer. J. Ag. Econ.* 68(1986):1196-1207.

Bourque, P. J. *The Washington State Input-Output Study for 1982.* Grad. School of Bus. Admin., Uni. of Wash., March, 1987.

Miller, R. E. and P. D. Blair. *Input-Output Analysis.* Englewood Cliffs, NJ: Prentice-Hall, 1985.

Pyatt, G. "Collapsing SAM's: A Technique with Particular Relevance for the Computation of Tax Incidence." *Review of Income and Wealth.* 1986.

Pyatt, G. and J. I. Round, eds. *Social Accounting Matrices: A Basis for Planning.* Washington, DC: World Bank, 1985.

Robinson, S. and D. Roland-Holst. *Modelling Structural Adjustment in the U.S. Economy: Macroeconomics in a Social Accounting Framework.* Work. Pap. No. 440, Dept. Ag. and Res. Econ., Uni. of California-Berkeley, 1987.

Table 14.1. Aggregate SAM for the United States, 1982 (Import Ridden)

	Activities			Value Added			Institutions/Households						Exogenous			Total
	Agric	Ag.Rel.	Other	Labor	Cap.	Ind.Tax	Labor	Prop.	Ent.	L40HH	M40HH	H20HH	Cap.Ac.	Gvt.	RoW	
	1	2	3	4	5	6	7	8	9	10	11	12	13	14	15	
Activities	---------- A ----------						---------- C ----------									
1 Agriculture	49.91	93.81	9.81	0	0	0	0	0	0	5.26	8.45	6.71	-0.22	8.28	19.41	201.42
2 Ag. related ind.	70.63	1119.77	442.84	0	0	0	0	0	0	388.57	691.97	633.89	40.55	451.88	97.33	3937.43
3 Other industries	7.97	452.87	644.77	0	0	0	0	0	0	44.89	102.2	102.95	374.53	190.32	231.68	2152.18
Value Added	---------- V ----------															
4 Labor	18.79	1314.26	531.18	0	0	0	0	0	0	0	0	0	0	0	0	1864.23
5 Capital	45.13	701.08	200.05	0	0	0	0	0	0	0	0	0	0	0	0	946.26
6 Indirect taxes	3.64	217.21	37.92	0	0	0	0	0	0	0	0	0	0	0	0	258.77
Insts/Hslds				---------- Y ----------			---------- H ----------									
7 Labor	0	0	0	1612.91	0	0	0	0	0	0	0	0	0	0	0	1612.91
8 Proprietors	0	0	0	0	111.51	0	0	0	0	0	0	0	0	0	0	111.51
9 Enterprises	0	0	0	0	834.77	0	0	0	0	0	0	0	0	53.25	0	888.02
10 Low 40% HH	0	0	0	0	0	0	145.4	9.72	80.61	0	0	0	0	205.12	-0.18	440.67
11 Medium 40% HH	0	0	0	0	0	0	742.07	33.92	133.37	0	0	0	0	107.9	-0.51	1016.75
12 High 20% HH	0	0	0	0	0	0	725.44	67.87	225.32	0	0	0	0	49.71	-0.48	1067.86
Exogenous																
13 Capital account	0	0	0	0	0	0	0	0	388.05	-18.76	56.86	97.4	0	-115.24	6.55	414.86
14 Government	0	0	0	251.31	0	258.76	0	0	60.66	20.71	156.47	226.9	0	179.51	-24.42	1129.90
15 Rest of World	5.35	38.42	285.63	0	0	0	0	0	0	0	0	0	0	0	0	329.40
Total	201.42	3937.42	2152.2	1864.22	946.28	258.76	1612.91	111.51	888.01	440.67	1015.95	1067.85	414.86	1130.7	329.38	

Row Totals May Not Equal Column Totals Due To Rounding

Table 14.2. Aggregate SAM for the United States, 1982 (Imports Purged)

	Activities			Value Added			Institutions/Households						Exogenous			Total
	Agric	Ag.Rel.	Other	Labor	Cap.	Ind.Tax	Labor	Prop.	Ent.	L40HH	M40HH	H20HH	Cap.Ac.	Gvt.	RoW	
	1	2	3	4	5	6	7	8	9	10	11	12	13	14	15	
Activities	---------- A ----------						---------------------- C ----------------------									
1 Agriculture	48.44	91.05	9.52	0	0	0	0	0	0	5.11	8.20	6.51	-0.21	8.04	19.41	196.07
2 Ag. related ind.	69.92	1108.56	438.41	0	0	0	0	0	0	384.68	685.05	627.55	40.14	447.36	97.33	3899.01
3 Other industries	6.78	385.52	548.88	0	0	0	0	0	0	38.21	87.00	87.64	318.83	162.01	231.68	1866.55
Value Added	---------- V ----------															
4 Labor	18.79	1314.26	531.18	0	0	0	0	0	0	0	0	0	0	0	0	1864.23
5 Capital	45.13	701.08	200.05	0	0	0	0	0	0	0	0	0	0	0	0	946.26
6 Indirect taxes	3.64	217.21	37.92	0	0	0	0	0	0	0	0	0	0	0	0	258.77
Insts/Hslds				---------- Y ----------			---------------------- H ----------------------									
7 Labor	0	0	0	1612.91	0	0	0	0	0	0	0	0	0	0	0	1612.91
8 Proprietors	0	0	0	0	111.5	0	0	0	0	0	0	0	0	0	0	111.50
9 Enterprises	0	0	0	0	834.77	0	0	0	0	0	0	0	0	53.25	0	888.02
10 Low 40% HH	0	0	0	0	0	0	145.4	9.72	80.61	0	0	0	0	205.12	-0.18	440.67
11 Medium 40% HH	0	0	0	0	0	0	742.07	33.92	133.37	0	0	0	0	107.9	-0.51	1016.75
12 High 20% HH	0	0	0	0	0	0	725.44	67.87	225.32	0	0	0	0	49.71	-0.48	1067.86
Exogenous																
13 Capital account	0	0	0	0	0	0	0	0	388.05	-18.76	56.86	97.4	0	-115.2	6.55	414.86
14 Government	0	0	0	251.31	0	258.76	0	0	60.66	20.71	156.47	226.9	0	179.51	-24.42	1129.90
15 Rest of World	3.36	81.31	100.61	0	0	0	0	0	0	10.72	22.37	21.85	56.10	33.07	0	329.40
Total	196.07	3899.00	1866.57	1864.22	946.27	258.76	1612.91	111.51	888.01	440.67	1015.95	1067.85	414.86	1130.73	329.38	

Row Totals May Not Equal Column Totals Due To Rounding

Table 14.3. Matrix of Normalized Expenditure Shares, S

	Agric	Ag.Rel.	Other	Labor	Cap.	Ind.Tax	Labor	Prop.	Ent.	L40HH	M40HH	H20HH	Cap.Ac.	Gvt.	RoW
	1	2	3	4	5	6	7	8	9	10	11	12	13	14	15
1 Agriculture	0.2471	0.0234	0.0051	0	0	0	0	0	0	0.0116	0.0081	0.0061	-0.0005	0.0071	0.0589
2 Ag. related ind.	0.3566	0.2843	0.2349	0	0	0	0	0	0	0.8729	0.6743	0.5877	0.0968	0.3956	0.2955
3 Other industries	0.0346	0.0989	0.2941	0	0	0	0	0	0	0.0867	0.0856	0.0821	0.7685	0.1433	0.7034
4 Labor	0.0958	0.3371	0.2846	0	0	0	0	0	0	0	0	0	0	0	0
5 Capital	0.2302	0.1798	0.1072	0	0	0	0	0	0	0	0	0	0	0	0
6 Indirect taxes	0.0186	0.0557	0.0203	0	0	0	0	0	0	0	0	0	0	0	0
7 Labor	0	0	0	0.8652	0	0	0	0	0	0	0	0	0	0	0
8 Proprietors	0	0	0	0	0.1178	0	0	0	0	0	0	0	0	0	0
9 Enterprises	0	0	0	0	0.8822	0	0	0	0	0	0	0	0	0.0471	0
10 Low 40% HH	0	0	0	0	0	0	0.0901	0.0872	0.0908	0	0	0	0	0.1814	-0.000
11 Medium 40% HH	0	0	0	0	0	0	0.4601	0.3042	0.1502	0	0	0	0	0.0954	-0.001
12 High 20% HH	0	0	0	0	0	0	0.4498	0.6086	0.2537	0	0	0	0	0.044	-0.001
13 Capital account	0	0	0	0	0	0	0	0	0.437	-0.042	0.056	0.0912	0	-0.101	0.0199
14 Government	0	0	0	0.1348	0	1	0	0	0.0683	0.047	0.154	0.2125	0	0.1588	-0.074
15 Rest of World	0.0171	0.0209	0.0539	0	0	0	0	0	0	0.0243	0.022	0.0205	0.1352	0.0292	0
16 Total	1	1	1	1	1	1	1	1	1	1	1	1	1	1	1

Table 14.4. The Multiplier Matrix, M

		Agric	Ag.Rel.	Other	Labor	Cap.	Ind.Tax	Labor	Prop.	Ent.	L40HH	M40HH	H20HH
		1	2	3	4	5	6	7	8	9	10	11	12
1	Agriculture	1.4051	0.1039	0.0817	0.0738	0.0477	0	0.0853	0.0833	0.043	0.114	0.0884	0.0763
2	Ag. related ind.	1.9797	2.8124	1.8197	1.7348	1.1225	0	2.0051	1.9631	1.0102	2.6358	2.0682	1.8142
3	Other industries	0.4915	0.5528	1.8288	0.4487	0.2895	0	0.5186	0.5095	0.2601	0.6468	0.5333	0.4779
4	Labor	0.9418	1.1152	1.1417	1.7195	0.4653	0	0.8316	0.8147	0.4187	1.0835	0.8574	0.7548
5	Capital	0.732	0.5888	0.542	0.377	1.2438	0	0.4358	0.4268	0.2194	0.5695	0.4494	0.395
6	Indirect taxes	0.1464	0.1698	0.14	0.1071	0.0693	1	0.1238	0.1213	0.0624	0.1621	0.1277	0.1122
7	Labor	0.8148	0.9649	0.9878	1.4877	0.4026	0	1.7195	0.7048	0.3622	0.9374	0.7418	0.6531
8	Proprietors	0.0863	0.0694	0.0639	0.0444	0.1466	0	0.0513	1.0503	0.0259	0.0671	0.053	0.0465
9	Enterprises	0.6458	0.5195	0.4782	0.3326	1.0973	0	0.3844	0.3765	1.1936	0.5024	0.3964	0.3485
10	Low 40% HH	0.1396	0.1402	0.138	0.1682	0.1487	0	0.1944	0.1893	0.1433	1.136	0.1075	0.0946
11	Medium 40% HH	0.4981	0.5431	0.5457	0.7479	0.3946	0	0.8645	0.7003	0.3538	0.5272	1.4169	0.367
12	High 20% HH	0.5829	0.608	0.6045	0.7806	0.5487	0	0.9022	1.0518	0.4815	0.5899	0.4665	1.4105

The Matrix of Exogenous Leakage Multipliers, SM

		Agric	Ag.Rel.	Other	Labor	Cap.	Ind.Tax	Labor	Prop.	Ent.	L40HH	M40HH	H20HH
		1	2	3	4	5	6	7	8	9	10	11	12
1	Capital account	0.3573	0.3069	0.2887	0.2512	0.5453	0	0.2904	0.2916	0.5792	0.2545	0.2905	0.2974
2	Government	0.5246	0.5751	0.5456	0.6506	0.3913	1	0.5962	0.597	0.3639	0.6024	0.5927	0.5984
3	Rest of World	0.1181	0.118	0.1657	0.0982	0.0634	0	0.1135	0.1114	0.057	0.1431	0.1168	0.1041
4	Total	1	1	1	1	1	1	1	1	1	1	1	1

Table 14.5a. Reduced Aggregate SAM, Household Sectors Combined

		Agric 1	Ag.Rel. 2	Other 3	Hslds 5	Cap.Ac. 7	Gvt. 8	RoW 9
1	Agriculture	48.44	91.05	9.52	19.82	-0.21	8.04	19.41
2	Ag. related ind.	69.92	1108.57	438.41	1697.28	40.14	447.36	97.33
3	Other industries	6.78	385.52	548.88	212.85	318.83	162.01	231.68
4	Households	41.27	1525.66	570.45	0.00	0.00	389.07	-1.17
7	Capital account	17.40	270.26	77.12	135.50	0.00	-91.97	6.55
8	Government	8.89	436.63	121.58	404.08	0.00	183.15	-24.42
9	Rest of World	3.36	81.31	100.61	54.94	56.10	33.07	0.00
	Total	196.07	3899.00	1866.57	2524.47	414.86	1130.73	329.38

Table 14.5b. Reduced Aggregate SAM, Normalized Expenditure Coefficients

		Agric 1	Ag.Rel. 2	Other 3	Hslds 5
1	Agriculture	0.2471	0.0234	0.0051	0.0079
2	Ag. related ind.	0.3566	0.2843	0.2349	0.6723
3	Other industries	0.0346	0.0989	0.2941	0.0843
4	Households	0.2105	0.3913	0.3056	0.0000
7	Capital account	0.0887	0.0693	0.0413	0.0537
8	Government	0.0454	0.1120	0.0651	0.1601
9	Rest of World	0.0171	0.0209	0.0539	0.0218
	Total	1	1	1	1

Table 14.5c. Full Multipliers, Hshld Sectors Combined

		Agric 1	Ag.Rel. 2	Other 3	Hslds 5
1	Agriculture	1.4078	0.1069	0.0847	0.0901
2	Ag. related ind.	2.0396	2.8879	1.8857	2.1099
3	Other industries	0.5038	0.5662	1.8423	0.5400
4	Households	1.2484	1.3217	1.3187	2.0096

Table 14.6a. Reduced Aggregate SAM, With Value Added

		Agric	Ag.Rel.	Other	Hslds	Total
		1	2	3	5	Val. Add.
1	Agriculture	48.44	91.05	9.52	19.82	
2	Ag. related ind.	69.92	1108.57	438.41	1697.28	
3	Other industries	6.78	385.52	548.88	212.85	
4	Value added	67.56	2232.55	769.15	0	3069.26
7	Capital account	0.00	0.00	0.00	135.50	
8	Government	0.00	0.00	0.00	404.08	
5	Rest of World	3.36	81.31	100.61	54.94	
	Totals Used	196.07	3899.00	1866.57	3069.26	
	True hsld total				2524.47	

Table 14.6b. Reduced Aggregate Sam, Expenditure Coefficients from Value Added

		Agric	Ag.Rel.	Other	Hslds
		1	2	3	5
1	Agriculture	0.2471	0.0234	0.0051	0.0065
2	Ag. related ind.	0.3566	0.2843	0.2349	0.5530
3	Other industries	0.0346	0.0989	0.2941	0.0693
4	Value added	0.3446	0.5726	0.4121	0.0000
7	Capital account	0.0000	0.0000	0.0000	0.0441
8	Government	0.0000	0.0000	0.0000	0.1317
5	Rest of World	0.0171	0.0209	0.0539	0.0179
	Total	1	1	1	0.82250

Table 14.6c. Full Multipliers from Value Added

		Agric	Ag.Rel.	Other	Hslds
		1	2	3	5
1	Agriculture	1.4393	0.1354	0.1090	0.0917
2	Ag. related ind.	2.7765	3.5468	2.4548	2.1495
3	Other industries	0.6923	0.7373	1.9880	0.5501
4	Value added	2.3710	2.3814	2.2623	2.4891

Table 14.6c (cont). Percentage of "True" Multipliers

		Agric	Ag.Rel.	Other	Hslds
		1	2	3	5
1	Agriculture	102.4	130.4	133.4	98.7
2	Ag. related ind.	140.2	126.1	134.9	98.9
3	Other industries	140.9	133.4	108.7	99.5
4	Value Added	194.3	184.4	175.6	110.5

Table 14.7a. Reduced Aggregate SAM, With Earnings

		Agric	Ag.Rel.	Other	Hslds	Total
		1	2	3	5	Earngs.
1	Agriculture	48.44	91.05	9.52	19.82	
2	Ag. related ind.	69.92	1108.57	438.41	1697.28	
3	Other industries	6.78	385.52	548.88	212.85	
4	Earnings	52.42	1325.35	598.09	0	1975.86
5	Other VA	15.14	907.20	171.06	0.00	
7	Capital account	0.00	0.00	0.00	135.50	
8	Government	0.00	0.00	0.00	404.08	
5	Rest of World	3.36	81.31	100.61	54.94	
	Total	196.07	3899	1866.57	2524.47	

Earnings: Labor VA plus Proprietors Earnings prorated thus: Ag. 30%, Ag Rel 10%, Other 60%

Table 14.7b. Reduced Aggregate SAM, Expenditure Coefficients from Earnings

		Agric	Ag.Rel.	Other	Hslds
		1	2	3	5
1	Agriculture	0.2471	0.0234	0.0051	0.0079
2	Ag. related ind.	0.3566	0.2843	0.2349	0.6723
3	Other industries	0.0346	0.0989	0.2941	0.0843
4	Earnings	0.2674	0.3399	0.3204	0.0000
5	Other VA	0.0772	0.2327	0.0916	0.0000
7	Capital account	0.0000	0.0000	0.0000	0.0537
8	Government	0.0000	0.0000	0.0000	0.1601
5	Rest of World	0.0171	0.0209	0.0539	0.0218
	Total	1	1	1	1

Table 14.7c. Full Multipliers from Earnings

		Agric	Ag.Rel.	Other	Hslds
		1	2	3	5
1	Agriculture	1.4064	0.0959	0.0794	0.0822
2	Ag. related ind.	2.0064	2.6210	1.7609	1.9264
3	Other industries	0.4953	0.5004	1.8104	0.4930
4	Earnings	1.2168	1.0769	1.1999	1.8348

Table 14.7c (cont). Percentage of "True" Multipliers

		Agric	Ag.Rel.	Other	Hslds
		1	2	3	5
1	Agriculture	100.1	92.3	97.2	88.5
2	Ag. related ind.	101.4	93.2	96.8	88.7
3	Other industries	100.8	90.5	99.0	89.2
4	Earnings	99.7	83.4	93.1	81.4

15

Computable General Equilibrium Analysis at the Regional Level

David S. Kraybill

Introduction

The topic of this chapter is a particular input-output (IO) extension known as the computable general equilibrium (CGE) model. Following the theme of this book, the chapter focuses on CGE analysis at the level of the subnational region. The second section of the chapter emphasizes the restricted nature of the general equilibrium embodied in the IO model. The third section describes the microeconomic structure of an interregional CGE model. The fourth section presents a brief overview of the macroeconomic assumptions underlying CGE analysis. Data requirements of CGE analysis are discussed in section five. Section six describes policy applications of regional-level CGE analysis.

Toward a More General Equilibrium

Equilibrium Methods of Analysis

A rudimentary form of general equilibrium theory underlies the static I-O model, while more complete concepts of general equilibrium are embodied in extensions of the model. This section emphasizes the conceptual continuum linking IO and CGE models. Central to an appreciation of this continuum is the concept of equilibrium.

The concept of equilibrium, widely utilized in economic theories, is particularly important in IO analysis and its extensions. Equilibrium implies a balance among key economic variables, such as income and expenditure. Hicks defines static equilibrium as a state in which all economic agents "are choosing those quantities, which out of the alternatives available to them, they prefer to produce and to consume" (p. 15).

General equilibrium analysis attempts to measure the extent to which an exogenous change in one sector affects levels of activity throughout the economy. The identification of effects induced by a particular exogenous change is simplified if a state of equilibrium is assumed to exist prior to the exogenous change. Equilibrium need not actually occur in the "real world" for the concept to be useful in economic analysis. Rather, the concept of equilibrium has analytic merit if key economic variables tend toward equilibrium and if actual values are reasonably close to equilibrium values.

Static equilibrium methods, such as IO and CGE, proceed in the following manner: (1) a benchmark (initial) equilibrium is computed and values of variables are recorded, (2) an exogenous change, typically associated with a specific set of policies, is introduced into the model, and (3) a counterfactual (final) equilibrium is computed. The difference between the initial and final values of variables is interpreted as the effect of the policy change.

Attributing the difference between the benchmark and counterfactual equilibria to the exogenous change is plausible under the following conditions: (1) all key variables affecting economic behavior are included in the analysis, and (2) the behavioral relationships among variables are specified correctly. If key economic variables are left out, the solution values calculated in the counterfactual case are spurious.

The measurement of policy impacts using IO models may be questioned on the grounds that numerous variables of importance in mainstream economic theories are excluded from the basic IO model. CGE models incorporate a larger set of variables identified as important in neoclassical resource allocation theory and in theories of aggregate income determination. If these variables are correctly identified and specified, CGE equilibrium is more general than IO equilibrium and CGE estimates of policy impacts are more reliable than IO estimates.

Assumptions of Input-Output Analysis

An elaborate system of simultaneous equations developed by Léon Walras and published in 1874 expressed mathematically the concept of equilibrium in the interaction of economic agents and markets. The principal merit of Walras' general equilibrium system was its formal elegance rather than its utility in application.

Leontief, in the input-output model, transformed the Walrasian system into a rough-and-ready tool that could be implemented with available data (Leontief).

The IO model, on the basis of several simplifying assumptions, trades theoretical refinement for empirical tractability (Dorfman). Simplifying assumptions of the original Leontief model include:

1. *Assumption One*: Demand equations for final goods are dropped.
2. *Assumption Two*: Supply equations for primary inputs are dropped.
3. *Assumption Three*: Relationships among all variables are linear.

Assumption one implies that incomes generated endogenously in production have no effect on final demand. In the circular flow of expenditures and income in an economy, assumption one suppresses demand shifts which would otherwise be induced by the flow of endogenous factor payments to households. These factor payments are deposited in sterile value-added accounts where they have no interaction with the demand-side of the economy. Most linear extensions of the basic IO model relax the first assumption by placing the labor receipts row and the household expenditures column inside the transactions matrix. IO models incorporating feedback from production to demand in this manner are referred to as "closed" IO models.[1]

Assumptions one and two, together, imply that the fundamental economic concepts of scarcity and efficiency are ignored in the Leontief system. Supply constraints on primary resources are assumed either nonexistent or nonbinding and, hence, opportunity costs of resource usage remain unchanged as output changes. Relative prices, the signaling mechanism for efficient resource allocation in neoclassical economic theory, play no allocative role in the IO model. The suppression of supply constraints and relative prices eliminates the channels of economic interaction most essential to the neoclassical theory of resource allocation. "Supply-side" IO models relax assumption two by incorporating factor-supply constraints (Giarratani). Equilibrium remains highly restricted in these models, however, since output shares remain constant. Both "closed" and "supply-side" IO models retain a serious limitation of the basic model: relative price effects and commodity and factor substitution are ignored.

Assumption three is implemented by utilizing simple linear functions, instead of the nonlinear functional forms of neoclassical theory. El-Hodiri and Nourzad (1988) show that Leontief technology does not rule out Cobb-Douglas-type substitutions, since the Cobb Douglas imposition of constant value of inputs (in price-times-quantity terms) is consistent with the fixed a_{ij} used in IO analysis. However, shifts in factor and product use cannot be decomposed to identify quantity changes separately from price changes. The effect of the linearity assumption is that relative price-induced substitutions in production and consumption are underidentified.[2]

The CGE model is characterized by a fuller general equilibrium than either the basic IO model or its linear extensions since all three of the IO assumptions are relaxed. Assumption one is relaxed by incorporating demand functions whose arguments are income and relative prices. Assumption two is relaxed by including supply constraints on primary inputs, labor and capital. Assumption three is relaxed through the utilization of nonlinear functional forms.

Microeconomic Foundations of Interregional CGE Analysis

This section describes the elements of a CGE model designed for analysis at the regional level. The approach described here is "bottom-up," implying that the behavior of economic agents is determined at the level of the region.[3] Furthermore, constraints on the behavior of economic agents are region specific. Income-expenditure feedback among regions is captured and interregional trade flows are determined endogenously.

Overview of a Regional-Level CGE Model

The economic accounts underlying CGE models are nearly identical to those used in "closed" IO models. The structure of these accounts is illustrated in Figure 1 from the viewpoint of a single region within an interregional system of economies.[4]

A use matrix, shown in the upper left-hand corner of Figure 1, expresses intermediate commodity requirements of each industrial sector in the region.[5] Industry requirements of labor and capital are given in the value-added matrix. The sum of intermediate and value-added inputs constitutes the value of industry outputs, which are converted to commodities in the make matrix.[6,7]

Value added is disaggregated in the lower left-hand corner of the diagram into accounts representing industry purchases of capital and labor, and taxes on labor, capital, and output. Industry purchases of capital and labor are translated into factor incomes in the bottom right-hand corner. Production taxes accrue to government accounts in the center of the diagram.

In addition to tax receipts from production, governments receive household income taxes. Governments also receive and pay intergovernmental transfers, shown in the center of the diagram.

Household incomes from factors of production are augmented by transfers from governments. An arrow connecting households (bottom right-hand corner) to personal consumption (top right-hand corner) links factor markets to product markets, completing the circular flow of economic activity in the region.

Final demands vectors are shown in the upper right-hand corner of Figure 1. These vectors include commodity demand levels for investment, net interregional exports, net international exports, commodity purchases by state and local governments, commodity purchases by the federal government, and personal consumption.[8]

In summary, regions are linked by both interregional commodity and income flows. Interregional commodity flows are real flows, potentially moving into and out of each region. Interregional income flows are nominal flows that include interregional commodity payments, interregional factor payments, transfers between the federal government and the state-local government, taxes paid to the federal government, and federal transfers to households.

Structure of Production

Production relationships, in most CGE models, are based on the assumption of separability of intermediate and primary inputs.[9] On this basis, profit-maximizing levels of production are then chosen by producers in a series of two stages. In the first stage of the production decision, industry output Z_{jr} is given by a Leontief production function of the form

$$(1) \quad Z_{jr} = \min \left[\frac{V_{jr}}{a_{vjr}}, \frac{X^{n}_{1jr}}{a_{1jr}}, \ldots, \frac{X^{N}_{njr}}{a_{njr}} \right], \qquad \begin{array}{l} j = 1,\ldots,n \\ r = 1,\ldots,k \end{array}$$

where V_{jr} is value-added used in the *jth* industry in the *rth* region and X^{N}_{ijr} is the amount of the *ith* intermediate input used by the *jth* industry in the *rth* region; a_{vjr} and a_{ijr} are technical coefficients expressing the amounts of value-added and intermediate inputs required per unit of Z_{jr}. In addition to being nonlinear, the sectoral production functions of the CGE model differ from those of the basic I-O model by the inclusion of a supply-constrained, value-added aggregate consisting of labor and capital.

The assumption of separability permits optimal combinations of labor and capital in the value-added aggregate V_{jr} to be determined, apart from the intermediate input decision, in a subfunction of equation 1. Value-added consists of labor and capital combined according to the following technological relationship:

$$(2) \quad V_{jr} = A^{V}_{jr} L_{jr}^{\alpha_{jr}} K_{jr}^{1-\alpha_{jr}}, \qquad \begin{array}{l} j = 1,\ldots,n \\ r = 1,\ldots,k \end{array}$$

where A^{V}_{jr} is a shift parameter and L_{jr} and K_{jr} are labor and capital employed in the *jth* industry in region r. The parameter α_{jr} determines the extent of substitution between labor and capital. Demand functions for labor and capital are derived by minimizing the firm's cost of obtaining its value-added requirements.

The nonlinear relationship between labor and capital has important theoretical implications: the product transformation curve maintains its general equilibrium property that the marginal rate of transformation equals the relative price ratio at profit-maximizing levels. This property, absent in the IO model, holds in the CGE model because equation 2 permits substitution between labor and capital (Vanek, pp. 133-35).[10]

Structure of Demand

The consumer is assumed to choose commodities in a series of three nested stages, shown in Figure 2.[11] In the first stage, the consumer in region s chooses among commodities X^1_{is}, differentiated by product type (e.g., clothes versus cars). In the second stage, the consumer obtains his chosen stage-one quantities of the *ith* commodity by cost-minimizing substitution between domestic production X^{2D}_{is}, and foreign production X^{2M}_{is} (e.g., United States versus all other countries). In the third stage, the consumer minimizes the cost of acquiring his optimal domestic quantities of each commodity by substituting between home-region production X^3_{iss} and other-region production X^3_{irs} (e.g., Georgia-produced clothes versus clothes produced in the rest of the United States).

At stages two and three, each of the stage-one commodities is differentiated by geographic origin. Production differentiation by origin, an assumption attributed to Armington, is widely utilized in empirical analysis of international trade. Consumption of commodities occurs only at level one, where the commodity X^1_{is} is a composite formed by combining the *ith* commodity from various geographic origins. Commodities at stage two represent inputs into the formation of X^1_{is}, while commodities at stage three represent inputs into the formation of X^{2D}_{is}. Demand for commodities at stages two and three is thus a derived demand.

Demand functions of several types, including those derived from Leontief, linear-expenditure (LES), and other demand systems, may be used to determine stage-one levels of commodity purchases. Following Armington, stage-two substitution between domestic and foreign sources is determined by utilizing the following constant-elasticity-of-substitution (CES) function:

$$(3) \quad X_{is} = \psi^1_{is}\left[\delta^1_{is}X_{is}^{2M^{-\rho^1_{is}}} + (1-\delta^1_{is})X_{is}^{2D^{-\rho^1_{is}}}\right]^{-\frac{1}{\rho^1_{is}}}, \qquad \begin{array}{l} i = 1,\ldots,n \\ s = 1,2 \end{array}$$

where ψ^1_{is} is a shift parameter, δ^1_{is} is a parameter expressing the share of imports in total consumption of the *ith* commodity in region s, and ρ^1_{is} is the CES substitution parameter.

By cost minimization subject to equation 3, the ratio of imported and domestic commodities consumed in region r is given by

$$(4) \quad \frac{X_{is}^{2M}}{X_{is}^{2D}} = \left[\frac{P_{is}^{2D}}{P_{is}^{2M}} \bullet \frac{\delta_{is}^{1}}{(1-\delta_{is}^{1})} \right]^{\frac{1}{1+\rho_{is}^{1}}}, \qquad \begin{array}{l} i = 1,\ldots,5 \\ s = 1,2 \end{array}$$

Stage three commodities are aggregated in an Armington function, mathematically identical to equation 3, except that the arguments of the function are replaced by X_{irs}^{3} and X_{iss}^{3}. From the stage-three Armington function, the ratio of X_{irs}^{3} to X_{iss}^{3} is derived in a manner similar to equation 4.

Just as in production, nonlinear modeling of commodity demand has important theoretical implications. At each of the three stages of demand, optimal quantities are determined by the efficiency condition that the marginal rate of substitution equals the relative price ratio.

The above discussion touches only the dominant microeconomic features of a CGE model. The neoclassical mechanisms of resource allocation emphasized here operate primarily through changes in relative prices. Aggregate income and aggregate expenditure also play important roles in economic change. To gain a proper view of the role of these aggregate variables in general equilibrium models, it is necessary to examine the underlying macroeconomic assumptions.

Macroeconomic Closure

Macroeconomic closure of a general equilibrium model has both an accounting dimension and a behavioral dimension. This section discusses both dimensions of closure for a simple closed economy (i.e., an economy without trade). The discussion then proceeds to the more complex problem of closure in an open economy model (i.e., an economy with trade).

Closed-Economy Closure

In an accounting sense, the macroeconomic closure of a general equilibrium model defines the aggregate variables included in the analysis. An economic account is then constructed for each aggregate variable. Just as in double-entry bookkeeping, a set of accounting identities is typically established to allow subsets of accounts to be compared with other accounts within the system to ensure consistency.

A necessary condition for accurate accounting is that aggregate income should equal aggregate expenditure. In the language of the national income and product accounts (NIPA), this condition is expressed as

$$(5) \quad Y \equiv E$$

where Y is income and E is expenditure.

In a closed economy without government, expenditure consists of consumption (C) and investment (I). By substitution, identity 5 may then be expressed as a relation between savings (S) and investment:

$$(6) \quad Y \equiv C + I$$
$$(7) \quad Y - C \equiv I$$
$$(8) \quad S \equiv I$$

In general equilibrium analysis, a system of economic accounts is transformed into an economic model by (1) treating identity 8 as an equilibrium condition, and by (2) specifying the mechanism by which savings and investment are brought back into balance when exogenous forces disrupt the benchmark equilibrium. The behavioral dimension of macroeconomic closure of a general equilibrium model lies in the mechanism by which equilibrium in the savings-investment relationship is restored.

Macroeconomic closure is achieved in the basic IO model by setting investment exogenously and varying output until savings increases or decreases to this fixed level of investment. The I-O closure implies a somewhat simplistic theory of income determination that places the burden of macro-adjustment entirely on savings. Model results based on the IO closure may differ considerably from those obtained from a model with a more complex mechanism of macro adjustment.

In contrast to the IO closure, CGE closures are generally more closely related to dominant schools of macroeconomic thought. Instead of placing the burden of macro-adjustment solely on output and savings, CGE closures generally involve a richer set of variables.

In the CGE literature, three primary types of macroeconomic closure are identified: neoclassical, Keynesian, and Johansen.[12] In the neoclassical closure, the economy is assumed to be fully employed; investment adjusts endogenously along with savings until these two variables are equal. In the Keynesian closure, the economy is not necessarily fully employed. Adjustments to macro imbalances occur through changes in the level of employment, the price level, or both. In the Johansen closure, consumption is assumed to adjust until investment and savings are in balance; full employment is assumed to prevail.

Open-Economy Closure

Trade affects macroeconomic closure of a general equilibrium model in several ways. First, an account recording net flows between the home country and the rest-of-the-world (ROW) must now be added. This account is the balance of payments. Accounting consistency in the balance of payments requires that

$$(9) \quad S^F \equiv -B,$$

where S^F is the capital account surplus (net inflow of foreign savings) and B is the current account deficit.[13]

Second, accounting identity 8 is now altered to include net foreign savings. Thus, consistency in the economic accounts of an open-economy model requires that

$$(10) \quad S^D + S^F \equiv I,$$

where S^D is domestic savings.

The macro-adjustment mechanism that restores equilibrium between savings and investment is generally more complex in the open-economy case than in the closed-economy case because of the link between the balance of payments and the level of savings. In international models, the exchange rate and official reserves play key roles in balance of payments adjustments. These forces that equilibrate the balance of payments have an important effect on the level of savings and, hence, the macro-adjustment process. In interregional models with flows among regions within a nation, equivalents of the exchange rate and official reserves do not exist because a common currency is used in interregional transactions. Balance of payments deficits are resolved by different mechanisms at the domestic regional level than at the international level.

Closure at the Regional Level

In a domestic regional model, a complete accounting requires that a relation similar to identity 10 hold for each region.[14] In this case, however, S^F is defined as net monetary flows between the home region and all other (domestic and international) regions.

A starting point for the specification of a regional balance-of-payments constraint is found in the international trade literature concerning optimal currency areas. Mundell points out that factors of production moving freely across regional boundaries serve to reduce regional balance of payments deficits. Suppose, for example, that a decline in relative competitiveness, caused by falling labor productivity in the home region, reduces demand for the region's exports and worsens the regional balance of payments (Dow, p. 177). The declining marginal productivity of labor will lower the regional wage rate and induce an outmigration of labor, raising the marginal product of labor in the home region and lowering it in the other region. As relative factor productivity rises in the home region, exports will increase.

Alternatively, a balance of payments deficit may be induced in the home region by an increase in the growth of home-region income and, subsequently, an increase in the level of imports from other regions. Increased demand will raise the price of imports, increasing the marginal product of labor and, hence, the wage rate in regions where these imports are produced. In response to the relative wage rate change, labor will emigrate from the home region, lowering demand for imports and improving the regional balance of payments.

Migration of capital in response to marginal productivity differences, net transfers from the federal government, and flows of funds through the banking system may also relieve regional balance of payments deficits. The banking system may be particularly important, though implementation of regional closures incorporating flows of funds would be difficult because of inadequate data.

Interregional flows of income, expenditure, and factors may not necessarily restore deficits in the balance of payments. Recently, several British economists have proposed reformulations of the theory of cumulative causation in terms of regional balance of payments deficits (Thirlwall; Dow). Whether or not factor mobility leads to equilibration of the balance of payments for a particular region depends partly on whether marginal productivity curves are downward sloping as assumed in neoclassical theory. Theories of cumulative causation imply that marginal productivity curves may not be downward sloping. Rather, market forces may restore equilibrium in the balance of payments through income deflation rather than through price changes.

Important problems remain in the closure of regional CGE models. Unfortunately, the regional economics literature does not provide much assistance in resolving these problems. Regional economists have paid little attention to regional balance-of-payments issues, perhaps because of the lack of a signaling mechanism such as the exchange rate upon which to focus.

In the absence of a well-developed theory of the regional balance of payments, the regional modeler faces a difficult choice. The burden of adjustment to regional balance-of-payments deficits can be borne entirely by factors of production only if they move instantly. This assumption does not seem credible. Thus, an alternative mechanism must be designated to take care of short-run imbalances. Such a mechanism was devised by Jones and Whalley, who assume that the federal government finances any temporary regional balance-of-payments deficits. Though this approach is convenient, it appears to have little basis in fact.

This writer suggests an alternative closure that would endogenize the regional allocation of national savings in response to changes in regional trade balances. This closure is generally consistent the view that interregional flows of funds through the financial system play a key role in offsetting the gap between regional spending and regional income.

Data

Data sets for CGE analysis are generally built on the foundation of IO accounts produced by national statistical agencies, such as the Bureau of Economic Analysis (BEA) of the United States Department of Commerce. IO accounts consist of detailed industry, commodity, factor, and final-demand transactions refined so as to embody market-level equilibrium, as well as aggregate income-expenditure equilibrium. This data is then augmented by information from a variety of other sources: national income and product accounts (NIPA), official censuses, household surveys, and expert judgement. Numerous adjustments to the data are inevitably required to fit together disparate pieces that are assumed, in the aggregate, to add up to available control totals.

Implementation of a CGE model also requires data on model parameters. A procedure known as calibration is used in most CGE studies to determine parameter values (Mansur and Whalley, 1984). This procedure can be illustrated by a set of n model equations:

$$(11) \quad f(y, x; \beta) = 0 ,$$

where y is a vector of endogenous variables, x is a vector of exogenous variables, and β is a vector of unknown parameters (Lau, pp. 128-129). Vectors y and x are calculated from the benchmark data set. Calibration consists of solving equation (11) for the vector β. However, for most common functional forms, the system is underidentified with respect to this vector. For this reason, key parameters, such as elasticities, are specified exogenously using values from previous econometric studies or from expert judgement, so that the remaining parameters can be solved for determinately (Whalley and Trela, p. 86). In practice, the CGE modeler is faced with a dearth of adequate estimates for key parameters.

Regional-Level CGE Applications

Most existing CGE studies have been conducted at the national level.[15] Regional-level CGE applications have appeared only recently. Most of these studies utilize interregional models, though several have employed single-region models.[16] Issues analyzed in these studies include regional impacts of tariffs, tax incidence, energy policies, pollution abatement, international trade deficits, federal purchases, federal and state taxes, and transfer payments.

Liew used a six-region model of Australia to study the regional impacts of a 25% across-the-board increase in tariffs. Matrix inversion was used to solve a linearized version of the model. The tariff increases were found to stimulate import-competing industries and to induce cutbacks in export industries.

Mutti, Morgan, and Partridge used a six-region model of the United States to analyze the incidence of state-local business taxes. Under the assumption of perfect capital mobility, they concluded that capital in every region bears the average national burden of state-local business taxes, regardless of regional differences in the tax rate. Similar issues of regional-tax incidence were analyzed by Kimball and Harrison in a model with two regions, consisting of California and the rest of the United States (ROUS).

Jones and Whalley included six Canadian regions and one international region in a model used to assess a wide range of tax, expenditure, and energy policies initiated both by federal and by provincial governments. A noteworthy feature of this model is the endogenous determination of labor migration. Labor is assumed to have locational preferences and is therefore only partially mobile in response to regional wage-rate differentials.

Buckley used a two-region model of the United States to examine the regional impacts of federal policies to control acid rain. Capital owners were found to suffer losses in the industrial Midwest as a result of pollution controls; in the rest of the country, capital gained through greater productivity. Labor improved its position throughout the country as a result of acid-rain reduction.

Kraybill used a two-region model, including Virginia and the rest of the United States (ROUS), to analyze the regional impacts of the nation's trade deficit and changes in federal spending and taxes. Industrial sectors were chosen according to their relative importance in rural versus urban areas. Value-added was found to have grown more slowly in rural areas than in non-rural areas in Virginia as a result of the nation's trade deficit. This decline in value-added was mitigated, to some extent, by a rise in federal spending in the state.

Conclusion

In the past, a theoretical gap has existed between regional IO modeling and other approaches to regional analysis. This gap was perpetuated by crude theoretical foundations that made it difficult or impossible for the IO model to incorporate concepts and econometric results from regional studies related to labor supply and demand, migration, tax incidence, cost-of-living, and other issues. In contrast, CGE models employ variables, parameters and functional forms that are relatively similar to those of conventional microeconomic theory and to widely-used econometric specifications.

The conceptual advantages of CGE modeling are offset, however, to some extent by problems of implementation. Availability and reliability of data continues to constraint regional modeling of all types. The problem is more acute in CGE analysis than in IO analysis because a wider range of data is employed. In addition, the number of sectors that can be incorporated in a CGE models remains relatively small. CGE models employ a higher level of aggregation, in general, than IO models because of the complexity of solving large systems of nonlinear equations.[17]

Theoretical work is needed to support further regional CGE analyses. In particular, little attention has been directed to the problems of macroeconomic closure of regional models. Further development of appropriate macroeconomic closures for regional economic systems would contribute importantly not only to CGE modeling but to an understanding of regional economies in general.

Notes

[1] IO extensions that close the equation system with respect to households, while remaining linear, include a variety of models such as Type II and Type III multiplier models (Miernyk et al), economic-demographic models (Madden and Batey), and social accounting matrices (Pyatt and Round).

[2] Assumption three also implies constant returns to scale.

[3] In contrast, the "top down" approach determines the behavior of economic agents at the level of the nation. Constraints are national, rather than regional. Regional impacts are usually (though not necessarily) determined by utilizing fixed regional shares to disaggregate national output. Feedback among regions and interregional trade flows are ignored in the "top-down" approach.

[4] This diagram is adapted from Ballard et al.

[5] A use matrix shows amounts of intermediate commodity inputs required per unit of output in each industry.

[6] A make matrix shows the units of commodity output produced per unit of industry output. Make matrices reflect the fact that any particular commodity is often produced by more than one industry and that industries commonly produce more than one commodity.

[7] The term commodity, as used here, includes both goods and services.

[8] For simplicity, inventories are included in investment. State and local governments, represented in the diagram by a single entity, could be separated into distinct entities.

[9] In particular, weak separability between intermediate and primary inputs is assumed.

[10] However, the degree of curvature of the net production possibilities surface is greater than it would be if there were substitution among intermediate inputs.

[11] A two region system, consisting of regions r and s, is assumed in Figure 2. This diagram is adapted from Whalley and Trela.

[12] For discussions of CGE closures, see Rattsó, Lysy and Robinson. The neoclassical and Keynesian closures contain elements of the theories of aggregate analysis after which they are named, though neither closure incorporates the full range of concepts associated with these theories.

[13] The term "foreign savings" is used throughout the CGE literature to refer to the capital account, expressed in domestic currency.

[14] For the most part, regional I-O modelers have ignored problems of macroeconomic closure and the regional balance of payments. Recent work by Rose and Stevens on regional I-O closure is an exception.

[15] For reviews of national-level studies, see Shoven and Whalley, and Fretz, Srinivasan, and Whalley.

[16] Single-region CGE models are not discussed here. For examples, see Despotakis and Fisher, and Hertel.

[17] For a discussion of CGE solution methods, see Dervis, de Melo and Robinson, pp. 486-496, and Meeraus.

References

Armington, P. "A Theory of Demand for Products Distinguished by Place of Production." *IMF Staff Papers*, Vol. 16, No. 1, 1969, pp. 159-78.

Ballard, Charles L., Don Fullerton, John B. Shoven, and John Whalley. *A General Equilibrium Model for Tax Policy Analysis*. Chicago: University of Chicago Press, 1985.

Buckley, Patrick Henry. "An Interregional Computable General Equilibrium Model for the United States with Illustrations of Regional Impacts of Acid Rain Pollution Control Costs." Unpublished Ph.D. dissertation, Boston University, 1988.

Dervis, Kemal, Jaime de Melo, and Sherman Robinson. *General Equilibrium Models for Development Policy*. Cambridge: Cambridge University Press, 1984.

Dorfman, Robert. "The Nature and Significance of Input-Output." *The Review of Economics and Statistics*, Vol. 36, No. 2, 1954, pp. 121-133.

Dow, Sheila C. "The Capital Account and Regional Balance of Payments Problems." *Urban Studies*, Vol. 23, No. 3, 1986, pp.173-84.

El-Hodiri, Mohamed and Farrokh Nourzad. "A Note on Technology and Input Substitution." *Journal of Regional Science*, Vol. 28, No. 1, 1988, pp. 119-120.

Giarratani, Frank. "Application of an Industry Supply Model to Energy Issues." In William Miernyk, Frank Giarratani, and Charles Socher, (eds.), *Regional Impacts of Rising Energy Prices*. Cambridge, Mass.: Ballinger Publishing Co., 1978, pp. 89-102.

Hicks, John. *Capital and Growth*. London: Oxford University Press, 1965.

Jones, Rich, and John Whalley. "A Canadian Regional General Equilibrium Model and Some Applications." Working Paper No. 8621C, The Center for the Study of International Economic Relations, Department of Economics, The University of Western Ontario, London, Ontario, June, 1986.

Kimball, L.J. and Glenn W. Harrison. "General Equilibrium Analysis of Regional Fiscal Incidence." In Herbert E. Scarf and John B. Shoven (eds.), *Applied General Equilibrium Analysis*. Cambridge: Cambridge University Press, 1984.

Kraybill, David S. "A Computable General Equilibrium Analysis of Regional Impacts of Macro-Shocks in the 1980s." Unpublished Ph.D. dissertation, Virginia Polytechnic Institute and State University, 1988.

Lau, Lawrence J. "Comments on Numerical Specification of Applied General Equilibrium Models: Estimation, Calibration, and Data." In Herbert E. Scarf and John B. Shoven, (eds.), *Applied General Equilibrium Analysis*. Cambridge: Cambridge University Press, 1984.

Leontief, Wassily. *Input-Output Economics*. New York: Oxford University Press, 1986.

Liew, Leong H. "'Tops-Down' versus 'Bottoms-Up' Approaches to Regional Modeling." *Journal of Policy Modeling*, Vol. 6, No. 3, 1984, pp. 351-367.

Lysy, Frank J. "The Character of General Equilibrium Models Alternative Closures." Mimeo, Department of Economics, Johns Hopkins University, Baltimore, Maryland, 1983.

Madden, Moss and Peter W. Batey. "Linked Population and Economic Models: Some Methodological Issues in Forecasting, Analysis, and Policy Optimization." *Journal of Regional Science*, Vol. 23, No. 2, 1983, pp. 141-64.

Mansur, Ahsan and John Whalley. "Numerical Specification of Applied General Equilibrium Models: Estimation, Calibration, and Data." In Herbert E. Scarf (ed.), *Applied General Equilibrium Analysis*, Cambridge: Cambridge University Press, 1984.

Meeraus, Alexander. "An Algebraic Approach to Modeling." *Journal of Dynamics and Control*, Vol. 5, 1983, pp. 81-108.

Miernyk, William H., Ernest R. Bonner, John H. Chapman, Jr., and Kenneth Shellhammer. *Impact of the Space Program on a Local Economy: An Input-Output Analysis*. Morgantown: West Virginia University Library, 1967.

Mundell, R. A. "A Theory of Optimum Currency Areas." *American Economic Review*, Vol. 51, 1961, pp. 657-65.

Mutti, John, William Morgan, and Mark Partridge. "The Incidence of Regional Taxes in a General Equilibrium Framework." Mimeo. Grinnell College, Grinnell, Iowa, 1987.

Pyatt, Graham and Jeffrey I. Round. *Social Accounting Matrices: A Basis for Planning*. Washington, D.C.: World Bank, 1985.

Rattsø, Jorn. "Different Macroclosures of the Original Johansen Model and their Impact on Policy Evaluation." *Journal of Policy Making*, Vol. 4, No. 1, 1982, pp. 85-97.

Robinson Sherman. "Multisectoral Models of Developing Countries: A Survey." In H.B. Chenery and T.N. Srinivasan (eds.), *Handbook of Development Economics*, Vol II. Amsterdam: North-Holland, 1989.

Rose, Adam and Benjamin Stevens. Transboundary Income Flows in Regional Input-Output Models: Or How to Close a Regional IO Model Properly." Processed, Department of Mineral Resource Economics and Regional Research Institute, West Virginia University, Morgantown, West Virginia, 1988.

Shoven, John B. and John Whalley. "Applied General-Equilibrium Models of Taxation and International Trade: An Introduction and Survey." *Journal of Economic Literature*, Vol. 22, 1984, pp. 1007-1051.

Thirlwall, A.P. "Regional Problems are Balance-of-Payment Problems." *Regional Studies*, Vol. 14, 1980, pp. 419-25.

Vanek, Jaroslav. "Variable Factor Proportions and Inter-industry Flows in the Theory of International Trade." *Quarterly Journal of Economics*, Vol. 77, No. 1, 1963, pp. 129-42.

Walras, Léon. *Elements of Pure Economics*. Translated by W. Jaffé. Homewood, Illinois: Richard Irwin, Inc., 1954.

Whalley, John, and I. Trela. *Regional Aspects of Confederation*. Vol. 68, Royal Commission on the Economic Union and Development Prospects of Canada. Toronto: University of Toronto Press, 1986.

Figure 1: Overview of the Economy of a Single Region in an Interregional System

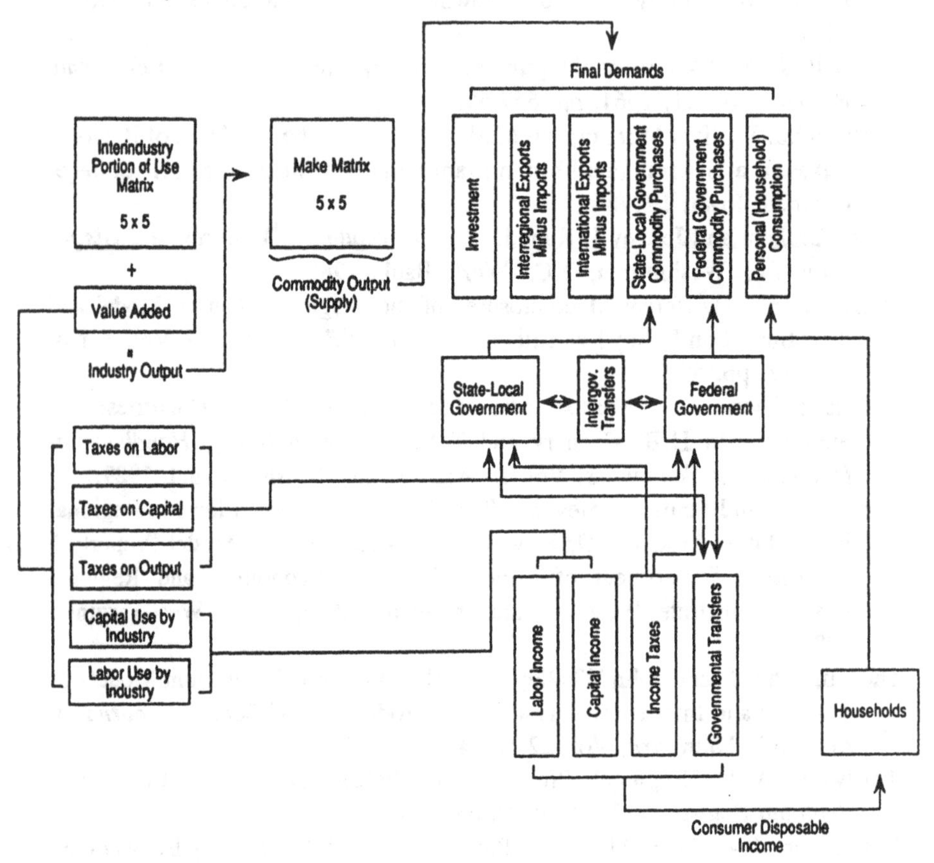

Figure 2: Hierarchical Structure of Demand in Region S

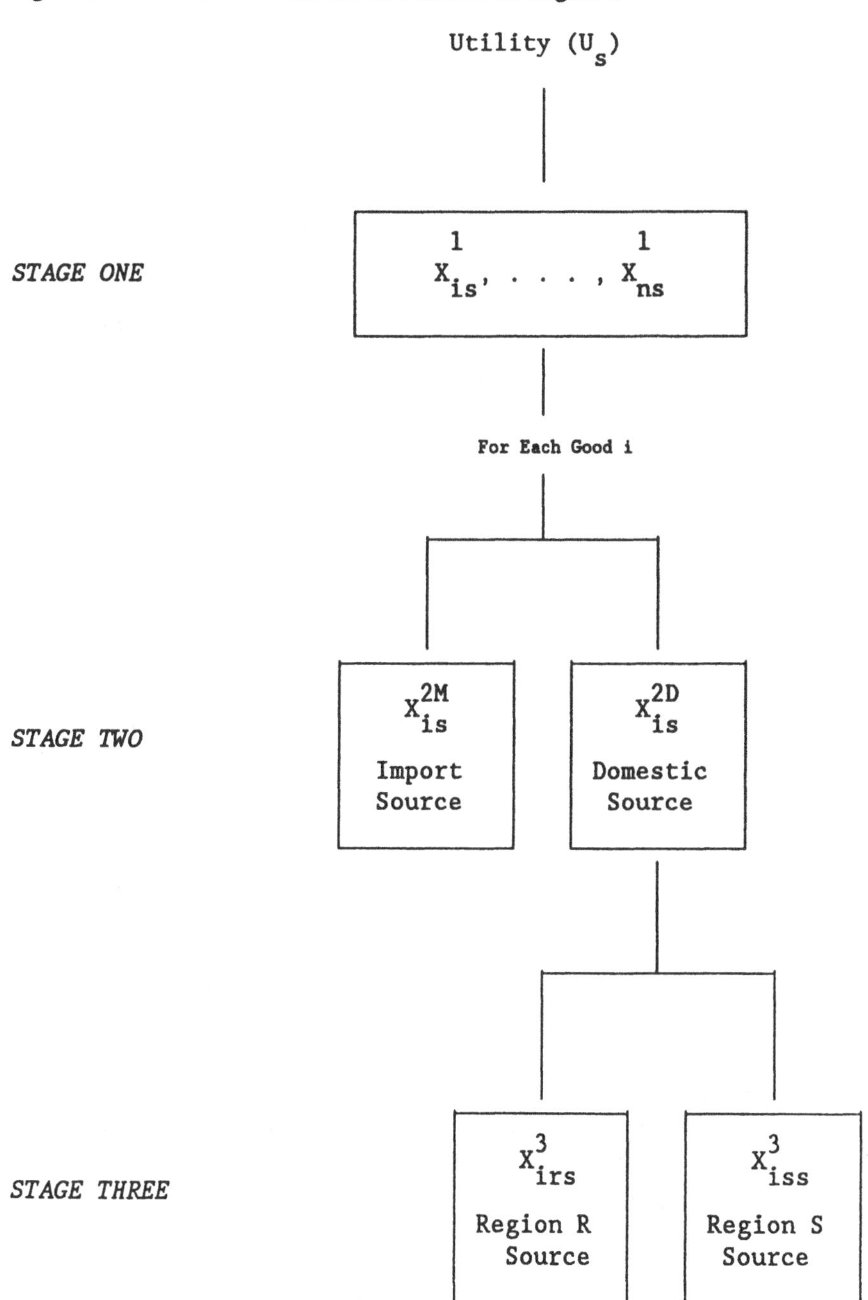

16

The Dynamics of Input-Output Introduction

Thomas G. Johnson

Introduction

Since the early experimentations with input-output analysis (Leontief, 1936, 1941), numerous variations have been offered. Of particular interest in this paper is a group of variations which add a temporal dimension to the otherwise static model. This paper develops a conceptual foundation for temporal input-output and reviews the experiences with the models.

Leontief introduced his dynamic version of input-output with the statement,

"A static theory derives the changes in the variables of a given system from the observed changes in the underlying structural relationships: dynamic theory goes further and shows how certain changes in the variables can be explained on the basis of fixed, i.e. invariant, structural characteristics of the system.

"Dynamic theory thus enables us to derive the empirical laws of change of a particular economy from information obtained through the observation of its structural characteristics at one single point of time. (1953, p. 53)"

Moore and Petersen state that a dynamic element is needed in input-output models "if the model is to be used for the analysis of regional business cycles, industrial growth and technological change, or for the computation of period (that is truncated) multipliers."

There are two major features which distinguish static from dynamic input-output models. First, as Leontief points out, the static model makes predictions given changes which are assumed to be exogenous (such as increased or decreased investment) while a dynamic model endogenizes these changes on the basis of more fundamental relationships (such as investment functions).

Second, the solution to the static balance equation is timeless. Final demand changes, require many rounds of transactions and a considerable amount of time to work themselves through the entire economy. In a static model the sectoral reaction paths are integrated, by the inverse matrix, into a single, total adjustment in each sector. The dynamic model, on the other hand, traces the paths of all endogenous variables as they react to changing final demand.

The static model is a static equilibrium device. Its solution indicates a set of circumstances in which there are no net forces of change acting upon the variables. This static equilibrium has a corresponding dynamic equilibrium in which, at each point in time, all variables are at levels in which there would be no net forces of change acting on them. Models describing these dynamic equilibrium conditions are incompletely dynamic since they do not describe the economy's movement in relation to its equilibrium values. A model which describes the movements of variables toward or away from their equilibrium levels in response to the net forces acting on them is known as a dynamic disequilibrium model. Equilibrium conditions are timeless, instantaneous functions of the actual economic conditions at a point in time whereas disequilibrium relationships involve lags, rates of change, and other temporal functions.

The most obvious way to add a temporal dimension to input-output is to calculate periodic solutions to the static model. This is often accomplished by embedding a static model in a system of temporal econometric equations (Tung, MacMillan and Framingham; Emerson, Lamphear and Atencio; Rubida; Bourque, Conway, and Howard; Treyz and Ehrlich). These models fall short of full dynamism because they do not trace the reaction path of the economy and they are void of disequilibrium relationships.

Moore and Peterson suggest a slightly more sophisticated approach in which the terms of the power series solution to the static model ($I + A + A^2 + A^3 + \ldots.$) would each represent a round of transactions and a fixed number of rounds, say 3, are assumed to occur each year. Johnson (1980), on the basis of experimentations with a dynamic input-output model assumed that 80 percent of the total impact from the static model would occur in the first year, 15 percent in the second, and 5 percent in the third year. Neither of these approaches are very satisfying, however. The following models are more truly dynamic. They follow directly from the static equilibrium input-output model. The following section begins with a static model and develops the dynamic model from it.

Static Equilibrium

The static input-output model is simply a system of structural identities:

$$(1)\quad X_t = AX_t + F_t \,,$$

where

- X is an n dimensional vector of industry output levels, (endogenous variables),
- A is an nxn matrix of coefficients relating the intersectoral purchases to the levels of sector output,
- F is an nx1 vector of final demand levels, that is, purchases by sectors other than industries (exogenous variables).

Static equilibrium exists when equation (1) is satisfied. It is usually more convenient to express this model in its reduced form, that is, where each endogenous variable is a function of exogenous variables only. Equation (1) may be solved as follows:

$$(2) \quad X_t = (I-A)^{-1}F_t$$

assuming (I-A) is non-singular.

Comparative static analysis with the static input-output model involves taking the total differentials of equations (1), as follows:

$$(3) \quad dX_t = (I-A)^{-1}dF_t \, .$$

For any i and j, i,j=1,n,

$$(4) \quad \partial x_1/\partial f_j = m_{ij},$$

where, m_{ij} is the i,j th element of $(I-A)^{-1}$.

The elements of the $(I-A)^{-1}$ matrix indicate the change in the endogenous variables in response to changes in the exogenous variables. The reference to time, t, indicates that the change in the equilibrium values is immediate. It does not, however, indicate that the value of the variable necessarily changes instantaneously by this amount. A dynamic model is needed to determine how rapidly the changes occur.

Dynamic Equilibrium

When the concept of time is explicitly incorporated into the static model, the concept of equilibrium changes significantly from the static equilibrium above. If each of the relationships above are expressed as functions of time, then the static balance equations become dynamic balance equations (conditions for dynamic equilibrium) as follows:

$$(5) \quad X(t) = AX(t) + F(t).$$

Dynamic equilibrium is defined as that set of conditions where, given the level of exogenous variables, there is no tendency for the endogenous variables to be at any other levels at that point in time. If the equilibrium is stable, variables have a tendency to decline (at that point in time) if they are above the equilibrium and a tendency to increase if they are below this point. This is not to suggest that all variables are constants over time in equilibrium. Only in a stationary state is this true. If the equilibrium is changing over time, a system is in dynamic equilibrium only if it is changing so as to continuously equal the equilibrium levels.

A stationary state is a special case of dynamic equilibrium. A stationary state is a unique dynamic equilibrium where the exogenous variables are constant. Other than the fact that this equilibrium is, at the same time, a dynamic equilibrium and a static equilibrium over time, the stationary state is largely an academic curiosity.

It is also clear that static equilibrium is a special case of dynamic equilibrium. Static equilibrium over any period does not assure dynamic equilibrium but dynamic equilibrium not only assures static equilibrium over the period but at every point in the period.

Another major difference between static and dynamic models lies in the distinction between endogenous and exogenous variables. Many variables which are exogenous in the static model are endogenous in the dynamic model. Leontief argued that in a dynamic input-output model, the final demand variable should be disaggregated into several component variables, some of which are best considered endogenous over time. In particular, capital purchases and changes in inventory are most appropriately viewed as functions of system variables. Leontief proposed the following model:

$$(6) \quad X(t) = AX(t) + Y(t) + B\dot{X}(t),$$

where, Y is redefined to exclude net investment in capital, B is an nxn matrix of capital-output coefficients reflecting how much capital stocks are needed by each sector to produce a dollar of output, and $\dot{X}(t)$ is an nx1 vector of changes in rates of output. This is a system of n linear differential equations which when solved give time paths for the n sector outputs. Leontief suggested a discrete version of the model in equation (5) as follows:

$$(7) \quad X_t = AX_t + Y_t + B(X_{t+1} - X_t).$$

Miernyk (Miernyk, Miernyk and Sears, Miernyk et al.) worked with models of the West Virginia Economy based on this model. A slightly more sophisticated dynamic equilibrium model would also recognize depreciation versus net investment, and inventory change. The following model incorporates these:

$$(8) \quad X(t) = AX(t) + Y(t) + BV(t) + \dot{N}(t),$$

where, Y is redefined to exclude gross investment in capital and changes in inventory, V is an nx1 vector of capital expenditures, B is an nxn matrix of capital-output coefficients reflecting how each sector spends its investment dollar, N is the level of inventories, and $\dot{N}$ is the time derivative of N.

Both of these models are equilibrium models and as such cannot adequately project the time paths of a real economy. Three aspects of the model are unrealistic. First they allow instantaneous and complete reversibility of investment. This allows unrealistically rapid accumulations and decumulations of capital (Leontief, 1970). Second, they don't provide for capacity constraints on output, or excess capacities. Third, there are no lags in the interindustry relations. Together these characteristics lead to reaction paths which are unrealistic and unstable. The following section describes a disequilibrium version of the above models which address these problems.

Dynamic Disequilibrium

Disequilibrium is basically a dynamic concept indicating that the levels of at least some variables are different than those indicated by the dynamic equilibrium equations. Disequilibrium models do not so much describe the levels of variables, but rather, how the variables change in response to changes in their equilibrium levels. Disequilibrium may be introduced by allowing for some potential divergence from the dynamic equilibrium conditions and those actually occurring at any time. In equation (8), any divergence from equilibrium levels of production will necessarily change the inventory levels. Changes in inventory then will equal the equilibrium change plus any change due to disequilibrium. This may be modeled by amending equation (8) as follows:

$$(9) \quad X(t) = AX(t) + Y(t) + BV(t) + N^{*}(t)\text{-}N(t) + E(t),$$

where, N^{*} is an nx1 vector of equilibrium levels of inventories, and E is an nx1 vector of rates of production over the equilibrium levels (either positive or negative). The difference, $N^{*}(t)$-$N(t)$, is the equilibrating change in inventories. Equilibrium inventories would usually be related to output in some proportionate fashion (although more complicated relationships are possible), for example,

(10) $N^*(t) = LX(t)$.

Equation (9) is an alternate balance equation, but one which is consistent with Equation (8). If $E(t) = 0$, and $\dot{N}(t) = N^*(t)\text{-}N(t)$, then the system is not only in static equilibrium but also dynamic equilibrium. This is not yet a disequilibrium model, but serves as the basis for one. Solving for E(t) gives,

(11) $E(t) = X(t) - [AX(t) + Y(t) + BV(t) + N^*(t)\text{-}N(t)]$.

If X(t) approaches its equilibrium levels at a rate proportional to the size of its gap between them, then

(12) $\dot{X}(t) = -\Phi E(t),\ \Phi > 0$,

where Φ is an nxn diagonal matrix of exponential lag coefficients, interpreted as the average number of transactional rounds in sector i each year--that is, its velocity. Equation (12) is the central relationship in this dynamic disequilibrium model. The model has the following properties:

1) it is stable since X(t) will increase (decrease) when production is less than (more than) consumption, and

2) when $E(t) = 0$, equation (5), (6), and (8) will be satisfied.

As pointed out above, capital investments must also be made endogenous in the dynamic model. Investment expenditures, V(t), arise for two reasons. First, they replace depreciated capital stocks, and second, they increase the capital stocks (the capacity) of the industry. Capital investment expenditures then are:

(13) $V(t) = DX^c(t) + \dot{X}^c(t)$.

If investment is less than that necessary for replacement, capital stocks and capacity decline. Capacity is a rather nebulous concept but an important one in dynamic models. Capacity is usually considered a function of the rate of production relative to demand. If we introduce the concept of equilibrium capacity X^{c*}, then theory would predict that it is a function of output, for example,

(14) $X^{c*}(t) = CX(t)$,

where C is a vector of constants > 1.0 reflecting the desired level of excess capacity at equilibrium. Actual capacity then will tend toward equilibrium along some adjustment path. If this path is exponential then,

$$(15) \quad \dot{X}^c(t) = K[CX(t) - X^c(t)],$$

where K is an nx1 vector of adjustment rates. By definition, output must not exceed capacity,

$$(16) \quad X(t) \leq X^c(t),$$

and gross investment, V (in equation 14) cannot be negative,

$$(17) \quad V(t) \geq 0.$$

Together, these equations are a complete system. In summary,

$$(18a) \quad \dot{X} = \Phi[AX + Y + B(DX^c + X^c) + LX - N - X],$$

$$(18b) \quad \dot{X}^c = K[CX - X^c], \text{ and}$$

$$(18c) \quad \dot{N} = X - (AX + Y + B(DX^c + X^c)),$$

subject to the constraints,

$$(19a) \quad X(t) \leq X^c(t), \text{ and}$$

$$(19b) \quad (t) \geq 0.$$

The system of equations (18), together with the constraints (19), form a dynamic input-output model. For any continuous time path of final demand F(t), initial conditions, and constants, time paths for outputs, X(t), capacity, $X^c(t)$, and investment, V(t), can be projected. While it is possible that these equations could be solved, yielding reduced form equations for the time path of each endogenous variable, the inequality constraints make this impractical. Fortunately, numerical methods and simulation techniques can greatly simplify this process.

As a means of demonstrating the essential features of an economic simulation model such as the one above, a simple one sector model was created. The model includes just three equations--one of each type in system of equations

Fortunately, numerical methods and simulation techniques can greatly simplify this process.

As a means of demonstrating the essential features of an economic simulation model such as the one above, a simple one sector model was created. The model includes just three equations--one of each type in system of equations (18). No constraints were included. Table 16.1 lists the values of the coefficients assumed.

Figure 1 shows the stationary state associated with the initial conditions in Table 16.1. Notice that at the levels of output and capacity, the static balance equation (1) and the dynamic balance equation (8) are satisfied.

Figure 2 illustrates the system's response to an instantaneous 20% increase in final demand at time 0. The system finally achieves the stationary state at an output of 2.474. The increase of .412 in response to an initial increase of .2 in final demand indicates that the system's output multiplier is about 2.06. This is somewhat more than the static multiplier of 2.0 because of the endogenous investment.

Figure 3 illustrates a more typical dynamic disequilibrium situation where the variables change toward a moving target. Figure 4 and 5 plot the predicted and equilibrium curves for capacity and output respectively. Since the system is in equilibrium at time zero, the curves start together but rapidly diverge. As the difference increases the rate at which the predicted curve approaches the equilibrium level increases.

Conclusions

The dynamic input-output model offers several advantages over the static model. First, it provides the analyst with a time path of impacts. Interpretation of the projections of the static model in a dynamic context is clearly wrong. The severity of this problem depends on the actual length of time required to exhaust the total impact of changes in final demand. Experimentation with the dynamic model above suggests that three or more years are required to substantially exhaust the impacts of increases in final demand which necessitate investments in sector capacity. Furthermore, it indicates that the timing of impacts varies from sector to sector (Johnson, 1986).

The dynamic input-output model predicts several variables which the static model does not including investment, capacity, capital stocks, and inventories. The time paths projected by the dynamic input-output model permit the analyst to make intertemporal comparison of variables and to calculate present values of streams of dollar values. A final advantage of the dynamic input-output model is that it allows the impact analysis of such dynamic final demands as seasonal or cyclical demands, phased expenditures, and time paths of investment and operating expenditures.

References

Bourque, P.J., Richard Conway, Jr., and C. T. Howard. "The Washington Projection and Simulation Model." Input-Output Series, University of Washington Graduate of Business Administration, September 1977.

Emerson, M. Jarvin, Charles F. Lamphear, and Leonard D. Atenciao. "Toward a Dynamic Regional Export Model." Annals of Regional Science3:2 (December 1969): 127-138.

Johnson, Thomas G. "A Dynamic Input-Output Model for Regional Impact Analysis." Unpublished PhD dissertation. Oregon State University, Corvallis, Oregon, 1979.

Johnson, Thomas G. "The Use of Simulation Techniques in Dynamic Input-Output Modeling." Simulation 41:3 (September 1983):93-101.

Johnson, Thomas G. "Simulation of Economic Systems," SP-86-16, Paper presented at the American Mathematical Society Meetings, New Orleans, January 7, 1986.

Johnson, Thomas G. "A Dynamic Input-Output Model for Small Regions." Review of Regional Studies 16:1 (Spring 1986): 14-23.

Leontief, Wassily. "Quantitative Input and Output Relations in the Economic System of the United States." Review of Economics and Statistics 19 (1936): 105-125.

Leontief, Wassily. The Structure of the American Economy, 1919-1929. Cambridge, Massachusetts: Harvard University Press, 1941.

Leontief, Wassily; Hollis B. Chenery; Paul G. Clark; James S. Duesenberry; Allen R. Ferguson; Anne P. Grosse; Robert N. Grosse; Mathilda Holzman; Walter Isard; and Helen Kistin. Studies in the Structure of the American Economy: Theoretical and Empirical Explorations in Input-Output Analysis. New York: Oxford University Press, 1953.

Leontief, Wassily. Input-Output Economics. New York: Oxford University Press, 1963.

Miernyk, William H. "The West Virginia Dynamic Model and its Implications," Growth and Change (April 1970): 27-32.

Meirnyk, W. H., Kenneth L. Shellhammer, Douglas M. Brown, Ronald L. Coccari, Charles J. Gallagher, and Wesley H. Wineman. Simulating Regional Economic Development: An Interindustry Analysis of the West Virginia Economy. Lexington: D.C. Heath and Co., 1970.

Miernyk, W. H., and J.T. Sears. Air Pollution Abatement and Regional Economic Development: An Input-output analysis. Lexington, Massachusetts: D.C. Heath, 1974.

Moore, F. T. "Regional Economic Reaction Paths." American Economic Review XLV (May 1955): 133-55.

Moore, F. T. and James W. Petersen. "Regional Analysis: An Interindustry Model of Utah," The Review of Economics and Statistics XXXVII:4 (1955): 368-383.

Rubida, Kirk W. "The Design of Computer Simulation Experiments With an Econometric Model of A Regional Economy: The State of Colorado." The Annals of Regional Science XII:2 (July 1978): 41-53.

Sargent, Thomas J. Macroeconomic Theory. New York: Academic Press, 1979. Treyz, George I. and David J. Ehrlich. "The Regional Economic Forecasting and Simulation (REFS) Model." Amherst, Massachusetts: Regional Economic Models, Inc., March 10, 1982.

Tung, Fu-Lai, James A. MacMillan and Charles F. Framingham. A Dynamic Regional Model for Evaluating Resource Development Programs." American Journal of Agricultural Economics 58 (August 1976): 403-414.

Table 16.1 Assumed Values for Coefficients of Hypothetical Model

Coefficient or Initial Condition	Value
Xc0	3.092784
N_0	.2061856
X_0	2.061856
K	.5
C	1.5
A	.5
B	.1
D	.1
OI	5.0
L	.1

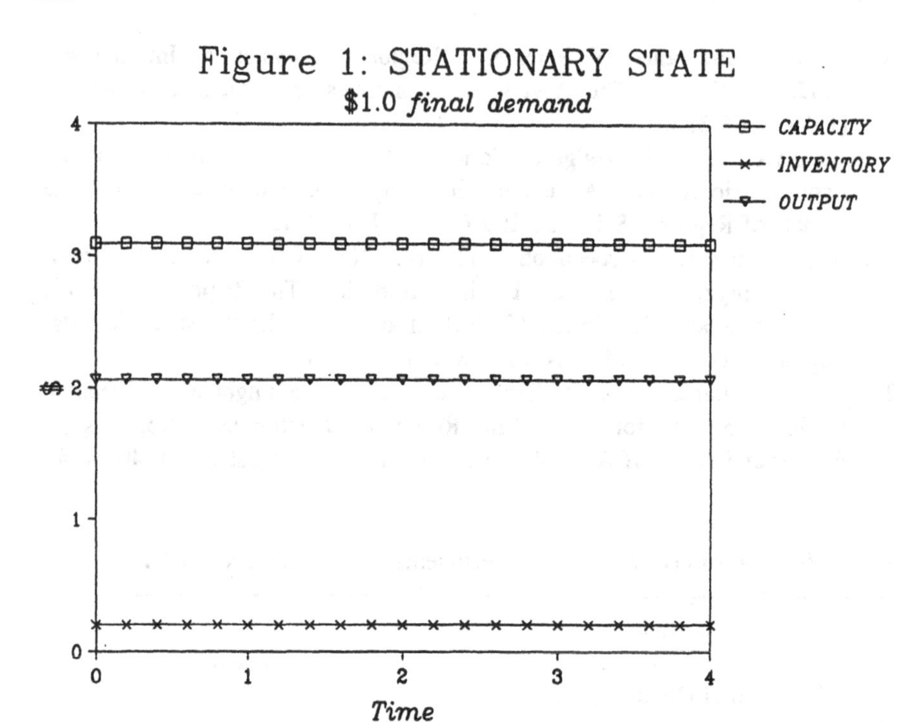

Figure 1: STATIONARY STATE

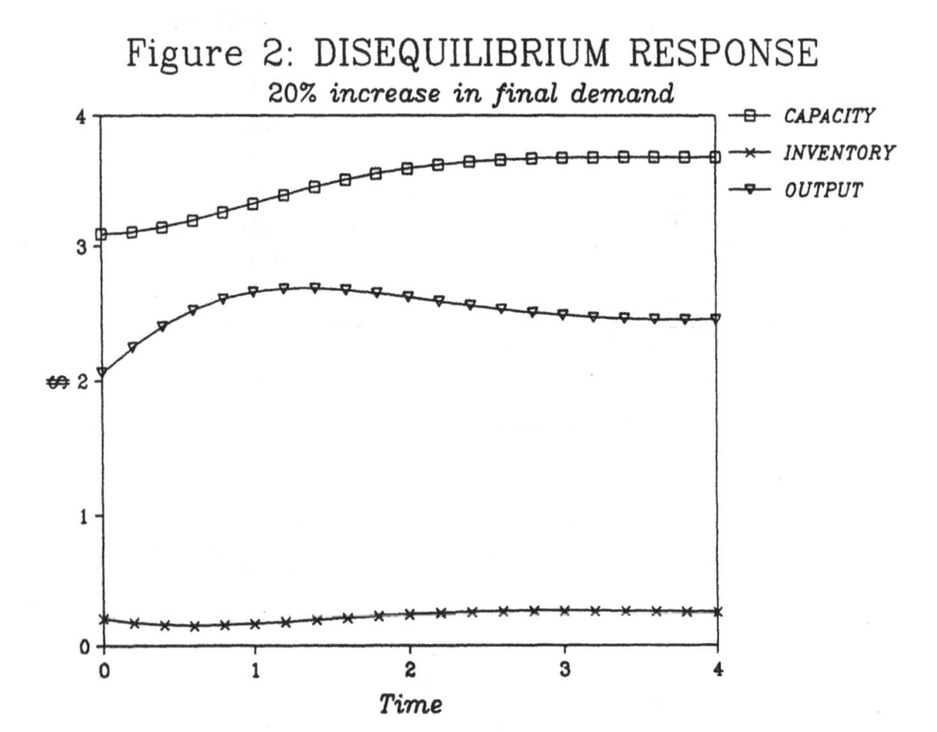

Figure 2: DISEQUILIBRIUM RESPONSE

Figure 3: DISEQUILIBRIUM RESPONSE

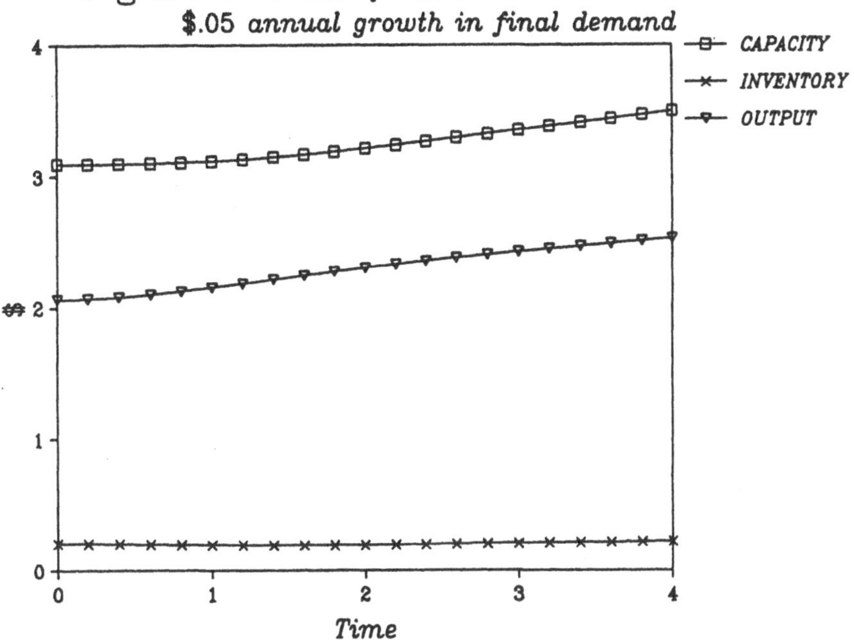

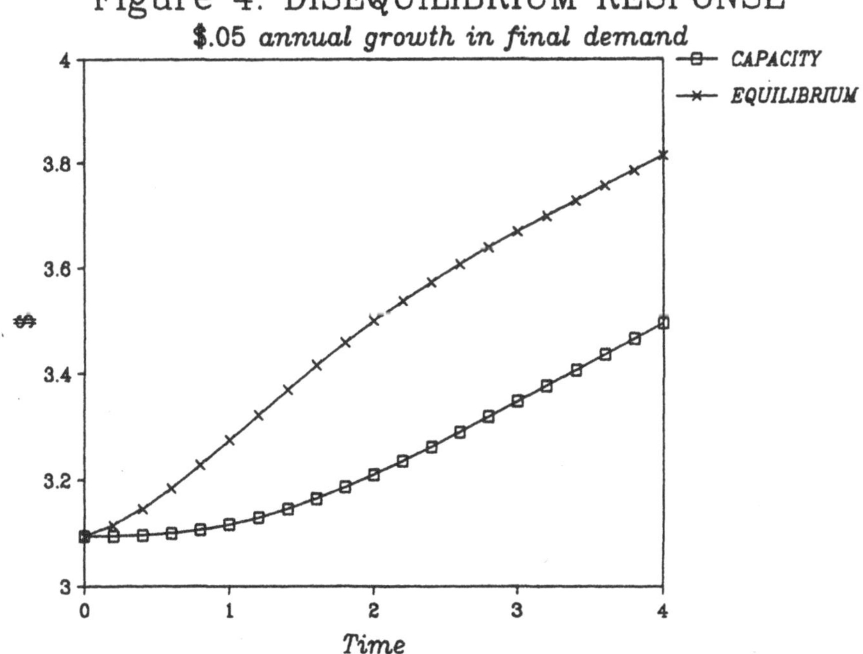

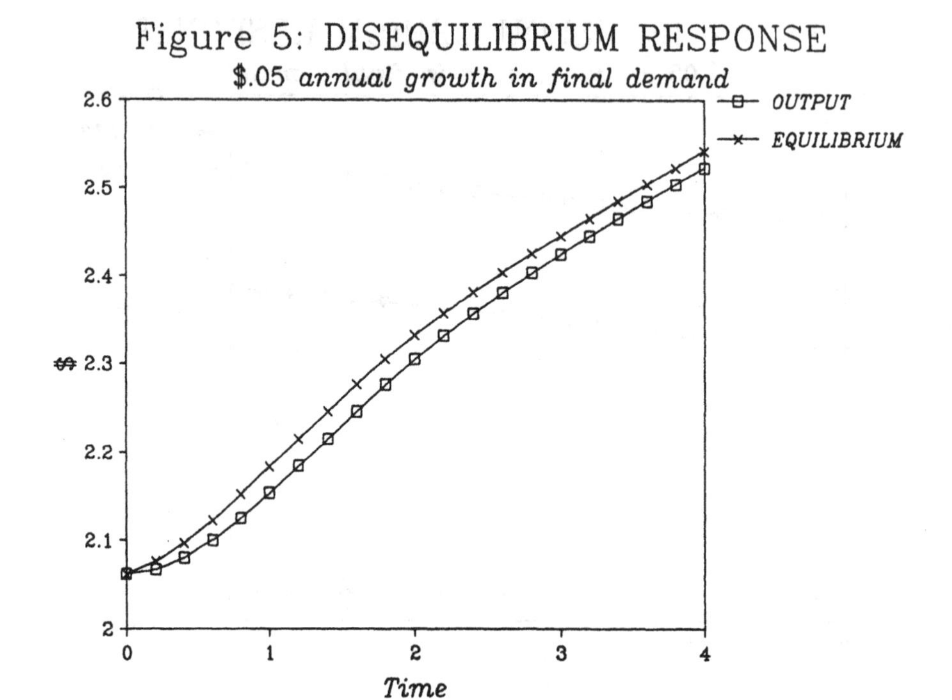
Figure 5: DISEQUILIBRIUM RESPONSE
$.05 annual growth in final demand
OUTPUT
EQUILIBRIUM
2.6
2.5
2.4
$ 2.3
2.2
2.1
2
0
1
2
3
4
Time